APPLIED MEDIA STUDIES

In the age of the maker movement, hackathons and do-it-yourself participatory culture, the boundaries between digital media theory and production have dissolved. Multidisciplinary humanities labs have sprung up around the globe, generating new forms of hands-on, critical, and creative work. The scholars, artists, and scientists behind these projects are inventing new ways of doing media studies teaching and research, developing innovative techniques through experimental practice. Featuring leading scholar-makers with years of experience creating applied media projects, this book presents behind-the-scenes stories, detailed case studies, and candid interviews with contributors. They describe projects such as reverse-engineering Spotify algorithms, building new mobile media networks in low-resource settings, hashtag activism, community-based locative storytelling app creation, collaborative and participatory design of medical media interfaces, and invention of new platforms for multimodal and transmedia storytelling. Readers will find practical advice and conceptual frameworks that prepare them to launch their own hands-on, participatory media projects using twenty-first-century tools and methods.

Contributors: Anne Balsamo, Heidi Rae Cooley, Jason Farman, Lindsay Graham, Daniel Grinberg, Eric Hoyt, Elizabeth Losh, Tara McPherson, Kirsten Ostherr, Lindsay Palmer, Lisa Parks, Bo Reimer, and Patrick Vonderau

Kirsten Ostherr is the Gladys Louise Fox Professor of English at Rice University in Houston, Texas. She is a media scholar and health researcher, and author of *Medical Visions: Producing the Patient through Film, Television and Imaging Technologies* (2013) and *Cinematic Prophylaxis: Globalization and Contagion in the Discourse of World Health* (2005), and co-editor of *Science/Animation* (2016). She is the director and co-founder of the Medical Futures Lab. Her current research is on information and communication technologies in medicine, patient narratives, and the role of simulation as a mediator between human and technological forms of medical expertise.

APPLIED MEDIA STUDIES

Edited by
Kirsten Ostherr

Routledge
Taylor & Francis Group

NEW YORK AND LONDON

First published 2018
by Routledge
711 Third Avenue, New York, NY 10017

and by Routledge
2 Park Square, Milton Park, Abingdon, Oxon OX14 4RN

Routledge is an imprint of the Taylor & Francis Group, an informa business

© 2018 Taylor & Francis

The right of the editor to be identified as the author of the editorial
material, and of the authors for their individual chapters, has been
asserted in accordance with sections 77 and 78 of the Copyright,
Designs and Patents Act 1988.

Library of Congress Cataloging-in-Publication Data
Names: Ostherr, Kirsten, 1970–editor.
Title: Applied media studies / edited by Kirsten Ostherr.
Description: New York: Routledge, Taylor & Francis Group, 2018. |
Includes bibliographical references.
Identifiers: LCCN 2017033806 | ISBN 9781138202481 (hardback) |
ISBN 9781138578265 (pbk.) | ISBN 9781315473857 (ebk)
Subjects: LCSH: Telematics. | Digital media. | Mass media.
Classification: LCC TK5105.6 .A65 2018 | DDC 384.3—dc23
LC record available at https://lccn.loc.gov/2017033806

ISBN: 978-1-138-20248-1 (hbk)
ISBN: 978-1-138-57826-5 (pbk)
ISBN: 978-1-315-47385-7 (ebk)

Typeset in Bembo
by codeMantra

For Benjamin and Theo.

CONTENTS

List of Figures   *x*
Acknowledgments   *xiii*
Notes on Contributors   *xvi*

Part I
Introduction   **1**

1 Applied Media Studies: Interventions for the Digitally
  Intermediated Age   3
  Kirsten Ostherr

2 Applied Media Studies and Digital Humanities:
  Technology, Textuality, Methodology   23
  Lindsay Graham

Part II
Foundations   **29**

3 Foundations of Applied Media Studies   31
  Kirsten Ostherr, Heidi Rae Cooley, Bo Reimer, Anne Balsamo,
  Patrick Vonderau, Elizabeth Losh, Eric Hoyt, Tara McPherson,
  and Jason Farman

4 From *Vectors* to Scalar: A Brief Primer for
  Applied Media Studies   48
  Tara McPherson

 5 The Medical Futures Lab: An Applied Media Studies
    Experiment in Digital Medical Humanities   60
    Kirsten Ostherr

Part III
Challenges   **79**

 6 Pleasures and Perils of Hands-On, Collaborative Work   81
    Kirsten Ostherr, Lisa Parks, Patrick Vonderau, Elizabeth Losh,
    Bo Reimer, Tara McPherson, Jason Farman, Heidi Rae Cooley,
    and Eric Hoyt

 7 Media Fieldwork: Critical Reflections on Collaborative
    ICT Research in Rural Zambia   97
    Lisa Parks, Lindsay Palmer, and Daniel Grinberg

 8 Rapid Response: DIY Curricula from FemTechNet
    to Crowd-Sourced Syllabi   117
    Elizabeth Losh

Part IV
Translation   **127**

 9 Transdisciplinary Collaboration and Translational
    Media-Making   129
    Kirsten Ostherr, Eric Hoyt, Tara McPherson, Bo Reimer,
    Lisa Parks, Jason Farman, Elizabeth Losh, Patrick Vonderau,
    and Heidi Rae Cooley

10 Collaborating across Differences: The AIDS
    Quilt Touch Project   141
    Anne Balsamo

11 The Time and Structure of Cross-College Collaboration:
    Developing a Shared Language and Practice   159
    Heidi Rae Cooley

Part V
Intervention   **179**

12 Unintended Consequences   181
    Kirsten Ostherr, Anne Balsamo, Jason Farman, Elizabeth
    Losh, Patrick Vonderau, Heidi Rae Cooley, Eric Hoyt, Tara
    McPherson, and Bo Reimer

13 Technology and Language, or, How to Apply Media
    Industries Research?   192
    Patrick Vonderau

14 Transforming the Urban Environment: Media
    Interventions, Accountability and Agonism   203
    Bo Reimer

Part VI
Infrastructure   **217**

15 Architectures of Sustainability   219
    Kirsten Ostherr, Tara McPherson, Heidi Rae Cooley,
    Patrick Vonderau, Lisa Parks, Jason Farman, Elizabeth Losh,
    Eric Hoyt, Anne Balsamo, and Bo Reimer

16 Building a Lantern and Keeping It Burning   238
    Eric Hoyt

Part VII
Conclusion   **251**

17 Conceptual Models and Helpful Thinkers   253
    Kirsten Ostherr, Jason Farman, Anne Balsamo, Patrick
    Vonderau, Elizabeth Losh, Bo Reimer, Heidi Rae Cooley, Tara
    McPherson, and Eric Hoyt

Index   *263*

FIGURES

5.1   Course Tumblr page for "Medicine in the Age of
Networked Intelligence," April 24, 2013   68

5.2   Team sepsis storyboard for "Medical Media Arts Lab," Spring 2017   70

5.3   Team e-Patient: Alex Lam, Obi Nwabueze, Mollie Ahn.
"Medical Media Arts Lab," Spring 2014   71

5.4   "Medicine in the Digital Age," MOOC poster. Spring 2015   73

7.1   A view of the dirt road into Macha's center, which people
use to move through the area by foot, bicycle, car, and mini-
bus each day   100

7.2   A view of the fire camp area adjacent to Macha's regional
hospital, where people wait for visiting hours and cook for
family members and friends   104

7.3   Students play outside at Macha Central School during recess   105

7.4   Consider Mudenda and Lindsay Palmer discuss our
interview with Macha Central School teachers   106

7.5   Project collaborators transfer interview footage from
flipcams to computers in a small room we rented near Macha
Girls School   107

7.6   Research team members worked closely to view and discuss
interview footage at the end of each day   107

7.7   Project collaborators viewed footage and translated
interviews from Chitonga to English   108

7.8   By walking along paths in the community, we often
encountered sites and people of interest, including this
massive mobile phone tower and its guard   112

10.1   Image of AIDS Memorial Quilt on the Mall of Washington,
DC, 1996   142

10.2   Image of Marvin Feldman panel   143
10.3   Image of eight panels that became the first block
       of the AIDS Quilt   144
10.4   Image of NAMES Project Foundation Quilt web search page   145
10.5   Image of Gert McMullin sewing a Quilt panel   147
10.6   Image of the Virtual Quilt. AIDS Quilt Touch archive   148
10.7   Image of visitors using the AIDS Quilt Touch Virtual Quilt
       Tabletop Browser. AIDS Quilt Touch archive   150
10.8   Close up image of the AIDS Quilt Touch Virtual Quilt
       Tabletop Browser. AIDS Quilt Touch archive   150
10.9   Image of visitor taking photograph of panels displayed on
       AIDS Quilt Touch Virtual Quilt Tabletop Browser. AIDS
       Quilt Touch archive   151
10.10 AIDS Quilt Touch launches June 27, 2012   152
10.11 Image of visitor using AIDS Quilt Touch Interactive Timeline   154
10.12 Image of AIDS Quilt Touch Indigogo campaign. AIDS
       Quilt Touch archive   156
11.1   Concept Mapping (Spring 2014)   163
11.2   Brainstorming Content and Goals (Spring 2016)   163
11.3   Group Work—Context (Spring 2015)   164
11.4   Group Diagrams (Spring 2015)   164
11.5   Group Diagrams (Spring 2015)   165
11.6   User Experience Diagram—*Ghosts of the Horseshoe* (Fall 2012)   165
11.7   User Experience Diagram—Ward One app (Spring 2015)   166
11.8   Student-generated Wireframe for the Ward One App
       (Spring 2016)   166
11.9   Slack Exchange (Spring 2016)   167
11.10 End-of-term Presentation and Demonstration—Question
       and Answer (Spring 2016)   168
11.11 Meeting with Ward One Organization and speaker Mattie
       Roberson (Spring 2015)   172
11.12 Meeting with Ward One Organization (Spring 2016)   173
11.13 Setting Up for Filming Formal Interview with Arthur Jones
       and Documentary Filmmaker Laura Kissel (Spring 2016)   174
11.14 Critical Interactive Students Travis Casey and Bonnie
       Harris-Lowe Prepare to Interview Former Ward One
       Resident Arthur Jones (Spring 2016)   175
13.1   An experiment using software tools to "sniff out" the
       communication between Spotify and its ad-supplying partners   199
14.1   The Parascope   207
14.2   Action shot from the Malmö City Symphony live
       performance. CC-BY Malmö City Symphony.
       http://citysymphony2009.blogspot.se/   209

14.3   Action shot from the Malmö City Symphony live
       performance. CC-BY Malmö City Symphony.
       http://citysymphony2009.blogspot.se/   210
14.4   Screenshot from the Malmö City Symphony live
       performance. CC-BY Malmö City Symphony.
       http://citysymphony2009.blogspot.se/   210
14.5   Screenshot from the Malmö City Symphony live
       performance. CC-BY Malmö City Symphony.
       http://citysymphony2009.blogspot.se/   211
16.1   This June 1927 issue of *Photoplay* was one of the first
       magazines digitized by the Media History Digital Library.
       It was scanned from the collection of the Pacific Film
       Archive, funded by an anonymous donation in memory
       of Carolyn Hauer. http://lantern.mediahist.org/catalog/
       photoplay3133movi_0741   241
16.2   The prototype of Lantern's first interface, developed in 2011   243
16.3   The redesigned version of Lantern, publicly launched in
       2013. http://lantern.mediahist.org   244
16.4   The Lantern team at the 2014 Society for Cinema and Media
       Studies Conference, Seattle, Washington. From left to right:
       Kit Hughes, Eric Hoyt, Tony Tran, Andy Myers, Joseph
       Pomp, Derek Long. Not pictured, but critical to the project's
       success were Carl Hagenmaier, Wendy Hagenmaier, Anne
       Helen Petersen, David Pierce, and Peter Sengstock   247
16.5   The Media History Digital Library's new homepage,
       launched in March 2017. http://mediahistoryproject.org   249

ACKNOWLEDGMENTS

I want to start by thanking the Andrew W. Mellon Foundation, in particular the New Directions Fellowship program, for giving me the opportunity to develop true bilingualism in my approach to applied media studies. In my proposal for the fellowship, I described my goal of getting a Master of Public Health degree as enabling me to "expand the applied humanism" of my work. The projects described in this book contribute to a new wave of applied, interdisciplinary humanities scholarship that seeks to intervene in, and thereby improve, the world that we live in. The Mellon Foundation's contribution to fostering the growth of this field has been invaluable to me and many others. Thank you.

The original idea for this book came from a workshop I chaired at the Society for Cinema and Media Studies conference in Montrèal, Canada (2015). The collaborative spirit of the speakers (Tara McPherson, Jason Farman, and Lisa Parks) and the audience in the room fueled a conversation that continued after the session ended, and eventually led to the creation of this book. I want to thank those original participants and all of the contributors to this collection for their brilliant, fearless, and creative dedication to making ideas into technologies people can think with. I am humbled and honored to be their editor and collaborator. Their work really defines why "applied media studies" matters.

Much of the inspiration for this book has come from my experience creating the Medical Futures Lab, where I do my own applied media studies work. My Medical Futures Lab co-founder and co-conspirator Bryan Vartabedian deserves special thanks for his energetic willingness to take part in a big experiment, his ability to adapt on the fly, and his tireless dedication to honing ideas into their most concise form. Thanks also to the other original participants who helped shape the Medical Futures Lab: Dave Thompson and the folks at ttweak,

Peter Killoran, Tom Cole, Olivia Banner, and Fred Trotter. I'm grateful to Carlos Monroy, Anthony Brandt, and Arnaud Chevalier, who brainstormed with me over coffee in the formative stages of the Medical Futures Lab. For their generosity in sharing resources on engineering design, I thank Matthew Wettergreen, Anne Saterbak, and Maria Oden. For their unique contributions to the Medical Media Arts Lab, I thank Tracy Volz, Allison Hunter, and Binata Mukhergee. Doctors Paul Checchia, Michael Fisch, and Monisha Arya generously served as exemplary "problem owners" early on, and helped us refine the clinician-facing parts of the design process. My thanks also go to the many students who have participated in Medical Futures Lab projects; their passionate commitment to changing the world makes all the hard work worthwhile.

Lindsay Graham has worked on this project with me almost from its inception, both in collecting and compiling our contributors' submissions, and in creating the companion Scalar site for the book. I thank her for her dedication, good humor, and professionalism. Many members of the Rice leadership believed in this project and helped along the way. As Dean of Humanities, Nicolas Shumway provided many forms of tangible and intangible support for the Medical Futures Lab. Paula Sanders and Cindy Farach-Carson provided much needed space for the Lab in the BioSciences Research Collaborative building. As Vice President for Strategic Initiatives and Digital Education, Caroline Levander's support has been instrumental to almost every project we've undertaken. In his role as Director of the Humanities Research Center, Farès el-Dahdah deserves special thanks for being an invaluable ally, source of intellectual collegiality, and inspired supporter of innovative projects at Rice, including those of the Medical Futures Lab.

Applied media projects are collaborative by nature, and I want to acknowledge three particular constellations that have energized and challenged me to expand the scope of my work. As a group focused on the intersection of people, technology, and design, Larry Chu and the Medicine X community based at Stanford have been an important and supportive test-bed for my applied medical humanities work. Several groups in the Image-Knowledge-*Gestaltung* cluster at Humboldt University in Berlin, Germany have provided valuable opportunities to expand on the work of the Medical Futures Lab in new clinical settings. I am grateful that Anna Roethe, Thomas Picht, Rebekka Lauer, Gunnar Hartmann, Kathrin Friedrich, and Moritz Queisner found the Medical Futures Lab online and pursued our collaboration across languages and time zones. I am also grateful to Kaisu Koski for reaching out to me from the Netherlands to collaborate on an applied medical media project. Our work together on standardized patients and clinical simulations has led to an exciting new kind of creative interventionist practice.

None of this would have happened without my home cheering section. My mother's courageous midlife return to school showed me the power of imagining new futures, and I am grateful to her for that, and for her unconditional love and support. My sister's inventive approach to outdoor experiential learning inspires me to bring those methods into scholarly contexts, and I always know I can count on her for encouragement. For the delight they bring into my world, I thank Benjamin and Theo, whose sense of play and ability to craft new worlds from Legos and cardboard boxes inspires me to aim toward a future where all of us find joy in work that makes the world a better, more just, more creative, and more humane place. Finally, my most affectionate appreciation goes to Peter, my brilliant sounding board, clinical translator, adventure companion, stand-up comedian, and inspired cook. The Medical Futures Lab, and all that has followed, never would have happened without you. Thank you.

NOTES ON CONTRIBUTORS

Anne Balsamo's work focuses on the cultural implications of emergent technologies. She is the author of the transmedia project, *Designing Culture: The Technological Imagination at Work* (2011) that examines the relationship between culture and technological innovation, with a particular focus on the role of the humanities in cultural innovation. She serves as the Dean of the School of Art, Technology, and Emerging Communication at the University of Texas at Dallas.

Heidi Rae Cooley is an associate professor of Media Arts at the University of South Carolina. Her book *Finding Augusta: Habits of Mobility and Governance in the Digital Era* and its digital supplement, *Augusta App*, won the Society for Cinema and Media Studies 2015 Anne Friedberg Innovative Scholarship Award. She is currently working on a second monograph tentatively titled, "Charles Sanders Peirce: Theorist for the Twenty-first Century."

Jason Farman is Associate Professor of American Studies, Director of the Design, Cultures & Creativity Program, and faculty member with the Human-Computer Interaction Lab at the University of Maryland, College Park. He is author of the book *Mobile Interface Theory: Embodied Space and Locative Media* (winner of the 2012 Book of the Year Award from the Association of Internet Researchers). He is the editor of the books *The Mobile Story* (2013) and *Foundations of Mobile Media Studies* (2015).

Lindsay Graham is a doctoral student in the Department of English at Rice University in Houston, Texas. Her research and pedagogical interests include Victorian poetry, gender and sexuality studies, and digital humanities. She is currently a Research Assistant with the collaborative "Resilient Networks to Support Inclusive Digital Humanities" project at George Washington University.

Daniel Grinberg is a PhD candidate in the Department of Film and Media Studies at the University of California, Santa Barbara. His research focuses on the intersections of documentary film, surveillance records, and the Freedom of Information Act. His writing has appeared in journals such as *Studies in Documentary Film*, *Jump Cut*, and *Spectator*, and is forthcoming in *Cinema Journal*, *Surveillance & Society*, and the *Journal of War and Culture Studies*.

Eric Hoyt is Assistant Professor of Communication Arts at the University of Wisconsin-Madison and co-director of the Media History Digital Library (http://mediahistoryproject.org), which has digitized nearly 2 million pages of books and periodicals related to film and broadcasting history. Hoyt is the lead developer of Lantern (http://lantern.mediahist.org) and one of the directors of Project Arclight (http://projectarclight.org). His publications include *Hollywood Vault: Film Libraries before Home Video* (2014) and *The Arclight Guidebook to Media History and the Digital Humanities* (2016), co-edited with Charles R. Acland.

Elizabeth Losh is a media theorist and digital rhetoric scholar, and associate professor of English and American Studies at the College of William and Mary. She is author of *Virtualpolitik: An Electronic History of Government Media-Making in a Time of War, Scandal, Disaster, Miscommunication, and Mistakes* (MIT, 2009) and *The War on Learning: Gaining Ground in the Digital University* (MIT, 2014), and co-author of *Understanding Rhetoric: A Graphic Guide to Writing* (Bedford/ St. Martin 2013). She is a founding member of FemTechNet.

Tara McPherson is Professor of Cinematic Arts and Director of the Sidney Harman Academy of Polymathic Studies at USC. She is author or editor of five books, including the collection *Transmedia Frictions* (2014) and a forthcoming monograph, *Feminist in a Software Lab*. She is the Founding Editor of *Vectors*, a principal investigator on the authoring platform, Scalar, and the recipient of numerous grants in support of these projects.

Lindsay Palmer is an assistant professor in the School of Journalism and Mass Communication at the University of Wisconsin-Madison. Her research focuses on global media ethics, looking specifically at global news reporting and the use of digital media technologies in transnational contexts. Palmer's work has appeared in *Television and New Media*, *Continuum*, *Critical Studies in Media Communication*, and *The International Journal of Cultural Studies*. She has two forthcoming books on the cultural labor of global news reporting in the twenty-first century.

Lisa Parks is Professor of Comparative Media Studies at Massachusetts Institute of Technology, where she directs the Global Media Technologies and Cultures Lab. She is author of *Cultures in Orbit: Satellites and the Televisual* and

Coverage: Vertical Mediation and the War on Terror, and has co-edited *Signal Traffic: Critical Studies of Media Infrastructures, Planet TV: A Global Television Reader,* and *Down to Earth: Satellite Technologies, Industries and Cultures.*

Bo Reimer is Professor of Media and Communication Studies at the School of Arts and Communication, Malmö University. He is one of the School's co-founders, as well as the founding director of *Medea,* a research lab dealing with collaborative media, design, and public engagement (http://medea.mah. se/). He is the author of *Collaborative Media. Production, Consumption, and Design Interventions* (The MIT Press 2013, with Jonas Löwgren), and *The Politics of Postmodernity* (Sage, 1999, with John R. Gibbins). He has also written *The Most Common of Practices. On Mass Media Use in Late Modernity* (Almqvist & Wiksell International, 1994). He sometimes works as a DJ.

Patrick Vonderau is Professor in Cinema Studies at the Department of Media Studies at Stockholm University. His most recent book publications include the co-authored *Spotify Teardown* (MIT Press, forthcoming 2017), and the co-edited *Films that Sell: Moving Images and Advertising* (Palgrave 2016), *Behind the Screen: Inside European Production Cultures* (Palgrave 2013), and *Moving Data: The iPhone and the Future of Media* (Columbia University Press, 2013). He is a co-editor of *Montage AV* and the *Media Industries Journal* and a co-founder of NECS-European Network for Cinema and Media Studies.

PART I
Introduction

1

APPLIED MEDIA STUDIES

Interventions for the Digitally Intermediated Age

Kirsten Ostherr

Introduction

In the era of "code for America," do-it-yourself (DIY) maker movements, hackathons, and STEM to STEAM, the field of media studies is ideally poised not only to analyze but also to intervene in the practices of media production and consumption that characterize our always-on, always-connected, screen-oriented lives. Doing so involves collaboration and translation across diverse modes of practice and fields of expertise. This work also requires inventing new strategies for conducting research, for presenting scholarly work, and for engaging with stakeholders beyond the academy. We call this emergent field of practice "applied media studies" for three key reasons. First, we want to emphasize a shared foundation in scholarship and methods that foreground the medium specificity of representational technologies as we explore how the digital intermediation of information and communication shapes meaning in the world. Second, we want to highlight the role that new kinds of hands-on, critical, and creative "applied" projects are playing in the evolution of scholarly practice in the field of media studies. Third, we want to explore and expand debates and methods at the intersection of humanistic media studies and applied sciences.

All of the projects described in this book represent collaborations among humanists, scientists, artists, and engineers, and in traversing those disciplinary boundaries, the concept of "applied" work in the humanities provokes comparisons with "applied sciences," raising important questions about the value and relevance of problem-oriented practice in relation to more abstract or theoretical work. This collection extends that dialogue by providing concrete examples of applied media work that offer new models for understanding humanistic knowledge formation as a productive field that can be "applied" to solving "real-world" problems, while also establishing feedback loops that bring new lines of inquiry back to more theoretical research. *Applied Media Studies* presents the insights and experience of media scholars who have forged innovative paths through this complex terrain, blending theory and praxis in applied projects that engage directly in the creation of new media infrastructures, communities, and texts.

Our goal in naming and defining the emergent field of applied media studies is to open up new approaches to research and teaching through screen-based technologies and media interfaces. Arising from the already multidisciplinary field of media studies, applied media studies adapts the methods for approaching traditional objects of inquiry in the field—that is, electronic, screen-based media—to respond to the demands of twenty-first-century research and teaching through critical experimentation on and with twenty-first-century tools. While older technologies of mediation such as cinema and television had fairly high barriers to entry, more recent iterations of these media enable forms of real-time participation that allow scholarship about media to become scholarship through media. The distinct material infrastructures of celluloid film and analogue television were certainly malleable, as vividly demonstrated by artists such as Stan Brakhage and Nam June Paik. But they were not nearly as malleable as the digital code upon which most of our mediated information, communication, and entertainment rest today.

In the contemporary era, more and more forms of visual, aural, and textual representation are digitally intermediated (that is, they appear, make meaning, and circulate through a complex mixture of digital platforms), perpetually redefining what comprises "screen-based media." Expanding on Bolter and Grusin's (1998) "remediation," scholarship that takes the act and infrastructure of mediation as its object of study is particularly well suited to translating theoretical and historical perspectives on representation into applied practice. As McPherson (2009) has argued, these practices of translation produce new kinds of mediated, multimodal technological interventions, and new kinds of social entanglements. Rheingold (2012) identified networked media creation as a unique sandbox for exploring life in the twenty-first century, noting, "When you start engaging in knowledge or media production, you tend to develop a much more sophisticated understanding of how knowledge and media is produced more generally" (84). As the case studies in this book attest, the work of translation, collaboration through difference, and experimental praxis has allowed the contributors to this collection to extend their scope and impact far beyond the boundaries of their academic settings by engaging directly in applied media studies projects with local and global communities.

Science and Humanities, Pure and Applied

By calling attention to the "applied" dimensions of the media studies projects presented in this collection, we gesture toward a long tradition of scholarly debate on the distinctions between "pure" and "applied" work that has profoundly shaped relations within and between the sciences and the humanities. In both domains, the "pure" end of the research spectrum has typically been valued more highly than the "applied" end. Yet in the era of digital intermediation, technological applications in fields such as statistics, applied mathematics, electrical engineering, and computer science have increased the value attributed to applied scientific work by society as a whole. One outcome of the digitization of everything from mechanical processes to interpersonal communication is that the output of applied science is experienced directly by industrialists and consumers alike. For example, discoveries in nanotechnology lead to production of faster and cheaper microprocessors that improve the quality and reduce the cost of new smartphones.

Another critically important outcome of ubiquitous digitization is that many previously elusive phenomena are being quantified and measured, as the conversion from analogue to digital entails the rendering of multidimensional and multisensorial signals into numerical form and binary code. As a result, value in American society has become inseparable from practices of quantification, and the digital era has become the era of the quantified self (Lupton 2016; Neff and Nafus 2016). In this context, the status of computational, quantitative applied sciences has been elevated, while the status of non-computational, non-quantitative humanities has declined. The "two cultures" famously described by C.P. Snow (1959/1961) as split into "polar groups" of scientists and literary intellectuals (also described more generally as "non-scientists"), are reproduced today as data-driven and non-data-driven cultures. But the projects described in *Applied Media Studies* show the result of collaborative work that could not be accomplished by scientists, technologists, or humanists working alone. Indeed, applied media studies provides a strong counter argument to the "two cultures" thesis because it shows that multidisciplinary efforts to reimagine the uses and meaning we make of digital networks, signals, and sensors result in substantive improvements to technology and society. A brief discussion of the origins and evolution of the "applied" versus "pure" debates will help frame the context for applied media studies today.

First coined by the English poet Samuel Taylor Coleridge, the term "applied science" appeared in his *Treatise on Method* (1817). However, according to historian of science Robert Bud (2012), Coleridge himself was translating a distinction established by the philosopher Immanuel Kant in his *Metaphysische Anfangsgründe der Naturwissenschaft* (*Metaphysical Foundations of Natural Science*) (1786). Kant distinguished between *"reine Wissenschaft* (pure science), which was based on a priori principles, and *angewandte Vernunfterkenntnis* (applied rational cognition)," based on empirical observation (Bud 2012: 538). Contemporary scholars see the Kantian distinction as shaping foundational ideas about the relative complexity, intellectual contribution, and cultural value of "pure" versus "applied" science (sometimes known simply as "technology"), whose legacy still shapes universities today.

Indeed, the distinction between "pure" and "applied" science has visibly shaped the funding, organization, and conduct of scientific research in the U.S. since World War II. In a 1961 *Technology and Culture* essay, James Feibleman defined "pure science" as "a method of investigating nature by the experimental method in an attempt to satisfy the need to know" (305). He contrasted this approach with "applied science," understood as "the use of pure science for some practical human purpose" (305). The instrumentalist connotations of "applied" work implicit in Feibleman's definition speak of the struggles for power, resources, and prestige that have driven heated debate over the meanings of these terms, both in the public sphere and in the academy (Bud 2012: 537). Scholars have shown that "the subordination of technological knowledge to scientific knowledge" led historians of "pure" science to exclude scholarship on the history of "applied" science or technology from conferences and publications in the postwar era (Alexander 2012: 518). Extending this analysis, historian Ruth Schwartz Cowan critiqued the implicit hierarchies associated with distinctions between "applied" technology and "pure" science in a lecture aptly titled, "Technology Is to Science as Female Is to Male" (1996). Summing up the implications of the analogy by noting, "Some say that science is theoretical and technology is practical; other people say that men are rational and women are practical" (576), Cowan's assessment of the gendered devaluing of the "practical" in this context provides a useful framework for interpreting the connotations of "applied" work in other contexts, including the humanities.

Debates about the relative merits of "applied" work have followed a tendentious path in the humanities. While the term "applied humanities" appears intermittently in print starting in the early twentieth century, the first extended discussion of the term in a scholarly text does not occur until 1974. In that year, Carnegie Mellon English professor Erwin Steinberg published an essay called, "Applied Humanities?" to argue that, in the face of increasing specialization at the "pure," abstract end of the spectrum of humanities scholarship, more humanists

> must bring to bear their professional expertise on contemporary problems. Doing so may frequently require not only that they move beyond their own narrow disciplines [...], but also that they address themselves to educated men and women outside of their own fields and sometimes outside of academia.
>
> *(445)*

Steinberg's rationale expressed a complex blend of conservatism, as he mourned the dwindling numbers of English professors dedicated to teaching traditional forms of literary appreciation and composition, and progressivism in his praise of emerging fields within English that studied unconventional topics in popular culture, such as television, detective novels, and westerns (446–47).

The essay goes on to cite several examples of the interdisciplinary, purposeful public engagement that the author celebrates under the heading of "applied humanism," including the founding of the journal *American Quarterly* in 1949, followed by the establishment in 1952 of the journal's scholarly organization, the American Studies Association, the creation in 1964 of the Centre for Contemporary Cultural Studies within the Department of English at the University of Birmingham, England, and the launch of the *Journal of Popular Culture* in the U.S. in 1967. Steinberg ultimately advocates teaching the insights of these novel fields as the primary vehicle for doing "applied humanities," though he also makes a brief call for what might be considered proto-digital humanities research, stating, "Indeed, I would urge at the blue sky end of the humanities continuum increased attention to some of the techniques that have proved so successful in the social sciences: statistical analysis, for example, and the use of the computer" (450).

It is perhaps ironic that Steinberg positions computational approaches to the humanities at the "pure" rather than the "applied" end of the spectrum where forty years later they found a home, since for many critics, practices such as text mining and distant reading (Moretti 2013) "fail the 'So what' test" (Schulz 2011) that defines knowledge-seeking at the heart of "pure" research. We will return to computational humanities below. Here, it is worth noting the unnamed backdrop to Steinberg's celebration of the democratizing forces that were bestowing scholarly legitimacy on topics previously considered the domain of "low" culture (Williams 1974; Levine 1988). The attention to popular culture and mass media in this era also opened up research on technology—the "low culture" of science—more generally, but this work developed alongside philosophical approaches at the "pure" end of the humanities spectrum, influenced by French theorist Jacques Derrida's (1967/1977) method of textual deconstruction. Steinberg critiques these more abstract methods, celebrating material culture at the expense of theory, but his view was not the dominant one in this era. The valorization of the "pure," theoretical work of deconstruction, poststructuralism, and postmodernism among literary scholars and many other humanists for decades signaled a distant stance toward the world outside of the academy, expressed through critiques of empirical claims about objective reality, including applied methods for approaching them.

The framework of "pure" versus "applied" has resurfaced periodically in debates about so-called crises in the humanities since Steinberg's publication in 1974. A piece from *The New York Times* called "The Applied Humanities: A Businesslike Approach" (Kolbert 1985) described the growth in programs dedicated to real-world application of humanities knowledge, or "bringing academic skills to bear on non-academic issues." That article emphasized the strategic turn toward "real-world" application of fields such as philosophy to help address existing problems as in, for example, ethicists helping to resolve biomedical conflicts in hospitals. Many of the sources interviewed for the article in *The New York Times* plainly admitted that the move toward "applied" work was driven by a shrinking job market and decreased overall support for many disciplines in the humanities. While such instrumentalist approaches to "applied" work leave many scholars concerned that the move from the "pure" to the "applied" end of the spectrum is a one-way path with potentially fatal, irrevocable consequences, others—particularly in the field of media studies—see the move as timely, politically necessary, and consistent with a wide range of scholarly methods.

For example, in an essay describing the response at the MIT Program in Comparative Media Studies to the attacks of 9/11, director of the program Henry Jenkins wrote of "applied humanism" as "the idea that insights from the humanities and social sciences need to be applied and tested at actual sites of media change." He noted that MIT already had applied physics and applied math, and argued, "It was time [MIT] had applied humanism. We challenged our students to do projects that had real-world impact and that confronted pragmatic challenges" (2004: 91). Jenkins does not dwell at length on the varied connotations of the term "applied." Instead, he focuses on describing the project that resulted from this mindset: the collaborative development of a website that both produced and analyzed media coverage following 9/11. Expanding on this model, the contributors to *Applied Media Studies* blend discussion of the conceptual frameworks and theoretical foundations that have guided their projects with descriptions of logistical challenges, technical hurdles, and social complexities that arise from moving their scholarly work into a real-world testing phase.

The call to "apply" critical media studies suggests a new role for critique that goes beyond the description and interpretation of cultural forms to demonstrate how critical perspectives might change or even improve existing circumstances in the world through direct intervention. While the purpose of this book and the projects described within it is not to provide a defense of the humanities, the work nonetheless demonstrates the power of bringing humanist methods of critical inquiry into the practice of technology development, interface design, and media production. At the same time, by claiming the modifier "applied" to describe the humanist media studies work in this collection, we put pressure on the long-standing but often misleading opposition between theoretical and practical work. In the years since the global financial crisis of 2008, much theoretical work in fields ranging from physics to literary studies has faced dwindling support from institutions of higher education responding to public pressures to narrow the field of intellectual inquiry to more instrumental objectives, achievable within a reduced time-to-degree framework, and oriented toward targeted vocational training (Brown 2016). This shortsighted approach to learning places little value on exploratory inquiry, creativity, aesthetic experience, and abstract forms of knowledge acquisition. It also fails to acknowledge several powerful currents in public culture and higher education, such as the concept of lifelong learning, the reorientation of research and teaching clusters around complex problems such as global warming or health disparities, and the growing recognition by many educational institutions and employers that some of the most desirable degrees or job applicants are those who have cultivated skills in critical and creative thinking, problem solving, ethical reasoning, and multimedia communication (Anders 2015; Henke et al. 2016; Shulman 2017; Deming 2017).

Countering the instrumentalist critique of applied work is the flourishing "STEM to STEAM" movement, which integrates science, technology, engineering, and math (STEM) with the arts and humanities. Pioneered by John Maeda when he was president of Rhode Island School of Design (2007–2013), the movement has gained significant momentum, attracting attention from policymakers, educators, and industry. In presenting a rationale for this approach, Maeda argued that "Art + Design are poised to transform our economy in the twenty-first century just as science and technology did in the last century" (STEM to STEAM 2017). In this spirit, a resolution was introduced to the U.S. House of Representatives (2013) affirming that integrating art and design into STEM policy "encourages innovation and economic growth in the United States" (H.RES.51, 113th Cong. 2013–2014). The STEM to STEAM movement has drawn attention because it responds directly to two critical challenges seen by many policymakers as vital to economic growth, social stability, and national security: the decline in the U.S. global ranking in STEM fields, coupled with high dropout rates and limited numbers of girls and historically underrepresented minorities going into STEM fields. The STEM to STEAM movement argues that integrating the arts and design is a powerful technique for attracting and retaining a stronger, more diverse technology workforce, thereby producing more innovative results at a national level (Maeda 2017).

The STEM to STEAM movement frames and makes visible an approach to humanistic academic engagement that extends beyond traditional structures of narrowly defined disciplines. While critics may reasonably object to the imperialistic resonances of a framing rationale rooted in aspirations of global STEM dominance, others might see in this approach a strategic form of instrumentalism. Maeda and other proponents of STEM to STEAM demonstrate insight in seeking policy-level affirmation of the vitality that the arts can bring to our data-driven world. They recognize that framing arts investment as a strategic national priority resonates with contemporary discourse on the role of creativity, flexible thinking, and risk taking—all attributes of artistic practice—in fostering innovation. Validation of this viewpoint through national and state policymaking is a powerful technique for ensuring ongoing resources and support for students, practitioners, and leaders of organizations. As a specific, concrete example of the STEM to STEAM approach, applied media studies shows the reciprocal benefits of bringing critical and creative practice together to intervene in the information and communication technologies that shape our world.

Critical Contexts

In its definition of "applied" science, the National Science Foundation describes "systematic study to gain knowledge or understanding necessary to determine the means by which a recognized and specific need may be met" (NSF 2010). The most pressing needs of our time—such as ending hunger, assuring access to clean water, and reducing armed conflict—affect humanity across the globe. These are social, human problems, but the digital platforms that connect populations through social and political engagement today are not being designed to meet our society's greatest "recognized and specific need[s]." The major global problems of the twenty-first century require multidisciplinary approaches, not narrowly technical solutions, and digital networks serve a core function in fostering dialogue across traditional boundaries. As daily life becomes ever more intermediated through the mobile platforms that are our constant companions, humanist media scholars are presented with opportunities to bring our scholarship directly to bear on large-scale public experiences through intervention in the design and implementation of digital interfaces.

The transformation of all forms of media production in the digital age invites—and perhaps demands—media scholars to reconsider how their work engages with the creation of media content that shapes the public sphere. Organizations such as the National Endowment for the Humanities and the Mellon Foundation have called for applied humanities work that tackles grand challenges. Doing so in the field of media studies requires critical reflection on the entanglement of technologies with their contexts of use. A key part of this consideration is the recognition that situated knowledges (Haraway 1988) ground the particular theoretical standpoints from which needs are identified and knowledge—including scientific or technical knowledge—is generated. This fundamental tenet of Feminist Science and Technology Studies plays a central role in much applied media studies work, shaping methodologies including collaborative participatory design and fieldwork, as well as "first-person" authorship practices that readers will see reflected in the writing style of this book.

Awareness of the situated dimensions of knowledge is particularly important because, in contemporary media ecosystems, we face many contradictory situations at the intersection of technology and human use. For example, the same media tools that were credited for democratizing access to the world's knowledge (Wikipedia) and connecting dissidents who sought to topple oppressive, authoritarian regimes (Twitter) have been critiqued for reproducing social hierarchies that exclude and exploit marginalized voices (Ratto and Boler 2014; Allen and Light 2015; Gordon and Mihailidis 2016). The need for more humanist engagement in technology development has been recognized by Silicon Valley observers who celebrate the founders of successful companies like LinkedIn and Slack that explicitly credit their humanities backgrounds as key fertilizers for their big, multimillion dollar ideas (Rowan 2012; Anders

2015). Feeding this recognition back into applied media studies does not mean designing pipelines to commercialization, but rather acknowledging and acting upon the translational potential of humanistic methods to intervene in existing problems at the intersection of human and technological needs.

The scholars in this collection launched their experiments from a foundation of expertise in media forms, histories, theories, aesthetics, infrastructures, and policies. By bringing their scholarly backgrounds into play with their work in making new forms of media interventions, the contributors to *Applied Media Studies* bring decades of experience exploring the mediation of cultural forms to the present digital media environment. The case studies and dialogues in this volume therefore draw from and extend a range of critical scholarly debates, expanding on science and technology studies and situated knowledge critiques to include scholarship on "useful" cinema, media industries, DIY and critical making, software studies, and digital humanities (DH).

In film and media studies, research on nontheatrical films produced by corporations, government agencies, and other organizations outside the entertainment industries has opened up a body of work exploring the uses of media to create deliberate, instrumental effects on audiences (Acland and Wasson 2011; Anderson and Dietrich 2012; Orgeron, Orgeron, and Streible 2012). Scholarship on these media forms, known variously as instructional, educational, useful, industrial, or sponsored films, highlights the frameworks that media studies can bring to bear on questions of how aesthetic, stylistic, narrative, formal, and historical dimensions of visual media design have been used to pursue specific effects in their viewers or users. Like "interaction" or "user experience" designers today, developers of instrumental films in the 1940s–1960s worked to create media that reshaped the perceptual experiences of users through their interaction with the screen. Situated between advertising and documentary genres, these films worked both to persuade and to explain, and the resulting mode of address often entailed an invitation to participate in a new worldview through engagement with new products or processes (Ostherr 2018). As Hediger and Vonderau (2009) note in their introduction to *Films that Work: Industrial Film and the Productivity of Media*, "industrial films are perhaps best understood as *interfaces* between discourses and forms of social and industrial organization" (11). Shifting this framing from instrumental to interventionist, applied media studies entails an integration of functionality and critique that allows for new understandings of mediation as critical making.

One contemporary site for work on industrial film is in the study of media industries at universities that pursue direct dialogue between academia and industry. Scholarship on media industries ranges from the historical work on nontheatrical film cited above to research on contemporary production practices, suggesting an obvious site for applied media studies work. However, scholars working in this field acknowledge the need for guidance on how to further develop their work as an "applied" practice. As Freeman (2017) has observed,

> While the study of media industries itself indicates a bridging between theory and practice—between the study of media forms and the pragmatics of media making—far less attention has been paid to what the study of media industries looks like as practice-based or practice-led research, or how collaboration between academia and the media industries actively shapes practice.
>
> *(n.p.)*

The tension between critique and cooperation with industry is an active topic of debate within applied media studies, both at the level of sponsored research and in cases where experiments with technology development intersect with start-up ecosystems.

Debates about the role of digital technologies in shaping the methodologies, objects of study, and dissemination techniques for humanist scholarship have led to the emergence of new research and teaching methods that embrace a "hands-on" philosophy. An exemplary project is *Critical Making* (2012), a DIY book project created and curated by Garnet Hertz, that "explores how hands-on productive work—making—can supplement and extend critical reflection on technology and society" (n.p.). Initially released as a 300-print run of hand-stapled and folded pages made on a "hacked photocopier," the project assembles 70 contributions from artists and academics whose work in some direct way engages and reflects upon the role of "making" in the contemporary era. Addressing an audience of "makers" outside the academy, Hertz calls on his readers to see the *Critical Making* project "as an appeal to the electronic DIY maker movement to be critically engaged with culture, history and society: after learning to use a 3-D printer, making an LED blink or using an Arduino, then what?" Hertz's call to engagement with academic critical frameworks reverses the common refrain that academics should engage more with the "real world"; instead he highlights the widespread appeal (both within and beyond the academy) of merging critical reflection with hands-on creative practice, demonstrating the synergistic value of these highly complementary—though often artificially separated—pursuits.

Media scholars have grappled with issues related to "DIY" media-making since the interactive web and the advent of hand-held digital cameras, smartphones, and YouTube radically transformed the practices of media-making that constitute the substantive focus of our field (Snickars and Vonderau 2010). In "What Happened before YouTube," Jenkins (2009) argues that fans have been producing their own media texts, infrastructures, and distribution networks since long before YouTube came along to make it easy for users to "broadcast themselves." Jenkins' work on participatory culture has spanned decades and demonstrates the deep histories of creative practice by non-specialists, outside formally defined media industries, that embody and inspire the approach taken by many of the scholars whose work makes up this collection.

Applied media studies practitioners have further integrated DIY and maker culture into the realm of humanistic media scholarship and practice. Löwgren and Reimer (2013) have demonstrated the enhanced insight that comes from bringing together different types of experts to approach a problem collaboratively as "expert amateurs." A key feature that DIY culture brings to academic practice is its agile approach to lab culture. As a method that does not depend on formal expertise or the expensive administrative and technical infrastructure that usually supports expert making in science labs or engineering fabrication sites, DIY praxis can rapidly adapt to new software, emerging formats, and current events in the world. Roued-Cunlife describes autodidactic and informal information sharing as key features of DIY culture, noting that this work "is rarely dependent on funding and thus has the potential to incorporate innovative solutions at a higher speed than we typically see in the humanities" (2016: 40). While humanities labs may be lower cost than their equipment-heavy counterparts in the applied sciences, the need for a lighter and more fluid interplay between humanistic work and relevant fields outside academia is well recognized by the authors in this collection.

A field adjacent to DIY making known as Critical Code Studies or Software Studies extends the tenets of critical making and the foundations of media theory to the analysis and building of video games and other forms of code. Ian Bogost's collection, *10 PRINT CHR$(205.5 + RND(1)); : GOTO 10* (2012), examines a single line of software code as a cultural text, drawing upon the editor's expertise as both video game developer and critical theorist. The project exposes the hidden assumptions embedded in a line of a popular programming language, thereby revealing the operations of the invisible algorithms that surround us within seemingly neutral technologies. As a design intervention itself, the book approaches its object through collaborative authorship of the chapters, each examining in hyper-close detail the implications of the coding framework. Another widely cited hands-on model is Manovich's collaborative work in the Software Studies Initiative (2008–2016), where computer scientists, art historians, and philosophers collaborate on experiments that blend software development with cultural analysis of large image-based datasets to analyze contemporary culture. Like the work of applied media studies, critical code studies do not merely "apply" a freestanding body of theory to some practical problem; rather, scholars follow a more iterative feedback loop methodology that approaches "making" from a theoretical standpoint and adapts as it learns from the experiment, fueling new questions and theoretical insights based on the hands-on work.

Much scholarship in the digital humanities shares the interest in learning by doing and hands-on critical thinking described here (Burdick et al. 2012; Gold 2012; Cohen and Scheinfeldt 2013; Gold and Klein 2016). Yet historically, few DH projects have originated in deep engagement with media studies, and consequently, the approach to "the digital" or to processes of mediation has often been focused on discussion of textual forms. While scholars from media studies have participated in debates in the digital humanities (Acland and Hoyt 2016), much of the debate in the formative years centered on issues related to the challenges and opportunities that arise when traditional text-based disciplines (such as literary studies) "go digital." Sayers (2018) characterizes the methodological tendencies of DH as rooted in disciplinary habits, noting, "With its prevalence in English departments and studies of literature and language, digital humanities frequently deems text its primary medium for both composition and analysis" (21).

However, as Graham (2018) notes in her contribution to this collection, the fields of digital humanities and applied media studies are not opposed, but rather complementary and evolving approaches to new forms of digital scholarship. Indeed, Ramsay (2013) has described certain kinds of DH work as "haptic engagement," to signal the high-level hermeneutic that emerges from the hands-on dimension of building that characterizes work ranging from text markup to map creation through GIS. The distinctions made in the pages of *Applied Media Studies* aim to extend the insights of both fields rather than define rigid, impermeable boundaries around either. Current developments in digital humanities also signal the importance of integrating disciplinary knowledge and methods, and Sayers' *Companion to Media Studies and Digital Humanities* (2018) provides an excellent starting point for pursuing the intersections of these fields.

Methods

The idea for this book came out of a workshop called "'Applied' Media Studies" that was held at the 2015 annual meeting of the Society for Cinema and Media Studies in Montréal, Canada. At that meeting, the workshop chair (Ostherr) moderated a discussion featuring several of the contributors to this volume: Tara McPherson, Jason Farman, and Lisa Parks. Many additional (future) contributors to this book were also in the room, and a lively discussion ensued. The rich experience and insight contained in the participants' descriptions of their collaborative, transdisciplinary projects was immediately evident, as was the need to capture the unique blend of conceptual framing and practical guidance that emerged. After the session ended, participants and audience members agreed that our field of practice—and the field of media studies as a whole— would benefit from a publication featuring the collective expertise in the room. In particular, many recognized the value in capturing the experimental perspectives that emerged as we discussed the issues we were currently grappling with. The recognition that tactical insights from the proverbial trenches might fade once projects had reached a "finished" stage led to a sense of urgency in capturing and disseminating the dialogue that the conference panel had provoked. This book is the result.

Applied Media Studies gathers the expertise gained through the contributors' experience in hands-on, collaborative media projects in order to preserve and share that knowledge and construct a foundation for future applied and theoretical work in this field. The work of doing applied media studies is inherently collaborative, and the writing in this volume reflects that quality through an approach to scholarly writing that de-emphasizes the single-authored monograph. Instead, the authorship methodology here is also collaborative. To produce six of the chapters in this book, contributors responded to a set of open-ended interview-style questions. As editor, I curated and synthesized their replies to provide a set of focused, diverse responses to the central questions raised in this book. Each multi-authored chapter is followed by detailed, single-authored case studies that explore the issues raised by the collaborative writing in greater depth.

The collaborative chapters in this book present a dialogue about the challenges and opportunities related to implementing applied media studies projects. Core issues include funding, sustainability, collaboration across disciplines, building conceptual models that draw on, extend and "apply" media theory, what it means to be a humanities-based Principal Investigator, how to build collaborative networks, and how to link these projects to teaching and research. Additional questions concern the complex, sometimes unintended consequences of intervening into practices that are more commonly studied from a distance, and the political and ethical implications of this work. *Applied Media Studies* demonstrates how new conceptual frameworks have emerged through the contributors' active participation in building new spaces, models and networks for engaged media studies.

The goal of this collection is to make the insights, challenges, and lessons learned by the contributors available to a broader audience of scholars interested in doing applied media studies work. At present, no resource exists to guide media scholars who want to engage in a hands-on, "start-up" approach to their work. Instead, many early experiments have taken place in relative isolation from each other, with limited public dialogue to support the development process. Indeed, the motivation for creating *Applied Media Studies* came from the contributors' direct experience with inventing new methods for engaging in experimental, interdisciplinary, collaborative work with "real-world" relevance and impact. Without exception, all of the contributors to this volume noted the lack of scholarly resources available to help guide their work as they redefined their humanist scholarship beyond the study of media, to include the creation of media interventions guided by theory and research. By publishing this collection and making additional, complementary material available online through a companion work on the open access Scalar platform (available at http://scalar. usc.edu/works/applied-media-studies/index), a community of scholars will be able to build on one another's experiences and extend these novel methods for bridging media theory and practice using twenty-first-century tools.

Chapter Outline and Case Studies

The structure of this collection moves between collaboratively authored chapters and single-authored case studies. Each section of the book addresses a core set of themes, with Section 1 providing the Introduction and a short piece by Lindsay Graham, "Applied Media Studies and Digital Humanities: Technology, Textuality, Methodology," that expands on the intersecting methods and affordances of the two fields of practice. Section 2, "Foundations," asks contributors to describe how and why they started doing applied media studies, and what, in their view, is the theoretical, historical, and/or political rationale for reimagining humanistic media studies as an applied practice. Two case studies follow the interview-style chapter. Tara McPherson's "From *Vectors* to Scalar: A Brief Primer for Applied Media Studies" explores the various ways in which collaborative, applied research challenges traditional models of humanities scholarship while also affording new ways to think about the role the humanities might play in the world. Drawing on over a decade's experience running a humanities-based digital publishing and software lab, McPherson elucidates the uneasy tensions produced by practice-based work within fields where the single-author monograph is the norm. Collaborative applied research often brings together teams of participants with diverse backgrounds and competencies whose scholarship and creative outputs do not neatly fit within the reward structures of research universities. McPherson addresses these challenges through a five-part primer for undertaking applied media studies. In the second case study, Ostherr describes the development of the Medical Futures Lab, an applied media studies project that emerged in response to a set of conceptual problems in medicine and media studies surrounding the role of technology in doctor-patient communication. Ostherr's essay demonstrates how applied media studies can lead from theoretical problems in human–computer interaction to practice-based answers in technology design, offering a model for a problem-based, field-specific humanities lab.

In the third section, "Challenges," contributors discuss the greatest pleasures and frustrations of doing applied media studies, describing how this work has impacted their teaching, research, publication, and service to the profession. This section also addresses the role of institutional setting, asking contributors to reflect on the ways that different contexts—such as urban research university, rural liberal arts college, elite private institution, or large, public institution—shape the kinds of work that are possible, and the varying resources and constraints that these settings impose. A case study follows by Lisa Parks, Lindsay Palmer, and Daniel Grinberg, called "Media Fieldwork: Critical Reflections on Collaborative ICT Research in Rural Zambia." This case study considers what it means to conduct media fieldwork by critically reflecting upon a collaborative research project in rural Zambia in 2012 and 2013. The chapter discusses what a media studies approach to fieldwork entails, what the affordances and limits of such an approach might be, and what kinds of unexpected results might emerge. Parks, Palmer, and Grinberg reflect on what happens when media scholars develop theory based on user research, and then design and build media resources in response. The second case study for Section 3 is Elizabeth Losh's "Rapid Response: DIY Curricula from FemTechNet to Crowd-Sourced Syllabi." Losh discusses applied media studies as a method for developing collaborative teaching practices that directly and rapidly respond to current social and political events. Through discussion of several examples, including the FemTechNet Distributed Open Collaborative Course, "The Selfie Course," #FergusonSyllabus, #Charlestonsyllabus, #Orlando Syllabus, and #BrexitSyllabus, Losh analyzes the logistics, politics, technical affordances, and broader scholarly and social implications of bringing hashtag activism into real and virtual classrooms.

Section 4, "Translation," asks contributors to address the core challenges of collaborating across divisions such as humanities and sciences, bridging academic and community practices, and translating between the diverse stakeholders involved in these projects. Participants discuss how they overcome the typical siloes of university structures, and how they translate between fields with radically different training, terminology, and theories of knowledge. Anne Balsamo's case study, "Collaborating across Differences: The AIDS Quilt Touch Project," discusses the process of creating a distributed cross-domain research team that included participants from the academy, the non-profit sector, government, and industry. Together, participants from seven different institutions across North America created digital experiences to augment the viewing of the AIDS Memorial Quilt. The essay elaborates on the process of assembling this team of collaborators, each of whom brought important expertise as well as unique, domain-specific (and biographical) background knowledges to the project: ways of knowing the world, ways of organizing workflow, ways of expressing, ways of negotiating participation. Balsamo's study describes the significance of some of those differences, and the process of negotiation required to collaborate across those differences. Heidi Rae Cooley's chapter, "The Time and Structure of Cross-College Collaboration: Developing a Shared Vocabulary and Practice," describes a team-taught course called "Critical Interactives." The course brings together undergraduate and graduate students from a variety of humanities disciplines, library and information science, and computer science to develop mobile applications for touchscreen devices that bring visibility to unacknowledged histories of racial and social inequity relevant to Columbia, SC and its flagship research university. Cooley explains how shared vocabulary and practice evolve complementarily over the course of a project's development, with examples from two "critical interactive" projects that move from the university into the community and back.

Section 5, "Intervention," addresses the complex, sometimes unintended results of intervening in practices that are more commonly studied from a distance, and the political and ethical implications of this work. The case studies for this section explore what happens when media scholars become actively involved in the reshaping of media experiences and infrastructures, thereby becoming part of the very processes they seek to critique. Contributors were asked about unintended ethical or legal issues, including Intellectual Property regulations they have had to address with respect to the open, collaborative work they created in online spaces. Patrick Vonderau's "Technology and Language, or, How to Apply Media Industries Research?" uses Spotify as a lens to explore the less obvious social, technical, and economic processes associated with "the digital," by both analyzing and intervening in the platform's functionality. Vonderau's case study is based on a multi-year project funded by the Swedish Research Council that engages in a reverse engineering of Spotify's algorithms, aggregation procedures, meta-data, and valuation strategies. The chapter investigates the processing of music as data by studying and intervening in the interaction between different actors involved at the 'back-end' or 'business-to-business' side of streaming services. In his case study, "Transforming the Urban Environment: Media Interventions, Accountability and Agonism," Bo Reimer asks what happens if you leave the safe environment of your university building and conduct research in the wild, starting processes in urban environments rather than just analyzing them. The answer, based on Reimer's experience of taking part in collaborative, interventionist research for the last fifteen years, explains the ethics and politics of carrying out such work.

Section 6, "Infrastructure," asks contributors how they have managed to attain the needed resources for their projects, and what kinds of institutional homes they have found to house them. Contributors also discuss how they cultivate the necessary team members as participants come and go, and what kinds of background, training, and mentoring of undergraduate and graduate students, as well as faculty members, is needed to do this kind of work (if any is needed at all). In the final case study for the book, "Building a Lantern and Keeping It Burning," Eric Hoyt describes and reflects upon the two years spent developing Lantern, from 2011 to 2013, and the two years following its official launch, from 2013 to 2015. Lantern was intended to provide a user-friendly search portal for the collections of the Media History Digital Library, but Hoyt also had the goal of using it to make an argument about the limited range of historical sources scholars typically use. The process of building Lantern spanned a transition from working at a kitchen table as a graduate student in California to working in a media department as a first-year professor in Wisconsin. Hoyt consistently depended upon the input of collaborators, but he found with equal consistency that he needed to learn new programming languages and open source technologies to keep the project moving forward. Yet, Hoyt notes that the work on Lantern will always remain incomplete—there are more scanned magazines to add to the index, sever breakdowns that require repair, and new users with new kinds of queries.

The Conclusion to *Applied Media Studies* builds on the dialogue that runs throughout the collaborative chapters in the book with an annotated compilation of the contributors' favorite resources for helping them, their colleagues, and their students engage in applied media studies work. Contributors were asked what kinds of conceptual models they have found helpful for extending and applying media theory as they move between making, writing, and teaching. Additional questions sought recommended articles, books, blogs, and Twitter streams, and finally, participants were asked to identify gaps in the field where they feel future research should be focused.

As more and more fields of scholarly practice are transformed by the digital turn, media scholars are uniquely positioned to help identify best practices in scholarship through interventionist media design. The contributors to this volume describe their experiences doing applied media work with practitioners from other fields, providing practical and conceptual guidance for engaging in hands-on projects with computer scientists, engineers, visual artists, medical doctors, and members of interested publics. As applied humanities labs continue to spring up in universities around the world, we hope that this book will serve as a valuable reference point for making those new approaches to scholarly practice viable over the long term.

References

Acland, Charles R., and Eric Hoyt, eds. (2016) *The Arclight Guidebook to Media History and the Digital Humanities*, Falmer: REFRAME/Project Arclight. http://projectarclight.org/book.

Acland, Charles R., and Haidee Wasson, eds. (2011) *Useful Cinema*, Durham, NC: Duke University Press.

Alexander, Jennifer Karns. (2012) "Thinking Again about Science in Technology," *Isis* 103(3): 518–26.

Allen, Danielle, and Jennifer S. Light, eds. (2015) *From Voice to Influence: Understanding Citizenship in a Digital Age*, Chicago, IL: University of Chicago Press.

Anders, George. (2015) "That 'Useless' Liberal Arts Degree Has Become Tech's Hottest Ticket." *Forbes*, August 17. www.forbes.com/sites/georgeanders/2015/07/29/liberal-arts-degree-tech/#22c0781d745d

Anderson, Nancy, and Michael R. Dietrich, eds. (2012) *The Educated Eye: Visual Culture and Pedagogy in the Life Sciences*, Dartmouth, NH: Dartmouth College Press.

Bogost, Ian. (2012) *10 PRINT CHR$(205.5 + RND(1));: GOTO 10*, Minneapolis, MN: University of Minnesota Press.

Bolter, Jay David, and Richard Grusin. (1998) *Remediation: Understanding New Media*, Cambridge, MA: MIT Press.

Brown, Sarah. (2016) "Bottom Line: How State Budget Cuts Affect Your Education," *The New York Times*, November 3. www.nytimes.com/2016/11/06/education/edlife/college-budgets-affect-your-education-but-its-not-all-bad-news.html.

Bud, Robert. (2012) "'Applied Science': A Phrase in Search of a Meaning," *Isis* 103(3): 537–45.

Burdick, Anne, Johanna Drucker, Peter Lunenfeld, Todd Presner, and Jeffrey Schnapp, eds. (2012) *Digital_Humanities*, Cambridge, MA: MIT Press.

Cohen, Daniel J., and Tom Scheinfeldt, eds. (2013) *Hacking the Academy: New Approaches to Scholarship and Teaching from Digital Humanities*, Ann Arbor, MI: University of Michigan Press.

Cowan, Ruth Schwartz. (1996) "Technology Is to Science as Female Is to Male: Musings on the History and Character of Our Discipline," *Technology and Culture* 37(3): 572–82.

Deming, David. (2017) "The Growing Importance of Social Skills in the Labor Market," *Quarterly Journal of Economics.* qjx022, https://doi.org/10.1093/qje/qjx022.

Derrida, Jacques. (1967/1977) *Of Grammatology*, Gayatri Chakravorty Spivak (trans.), Baltimore, MD: Johns Hopkins University Press.

Feibleman, James. (1961) "Pure Science, Applied Science, Technology, Engineering: An Attempt at Definitions," *Technology and Culture* 2(4): 305–17.

Freeman, Matthew. (2017) Conference Website. Journal of Media Practice and MeCCSA Practice Network Annual Symposium: "Practice and/as Media Industry Research." www.bathspa.ac.uk/liberal-arts/research/media-convergence-research-centre/meccsa-practice-network-symposium-2017/.

Gold, Matthew K., ed. (2012) *Debates in the Digital Humanities*, Minneapolis, MN: University of Minnesota Press.

Gold, Matthew K., and Lauren F. Klein, eds. (2016) *Debates in the Digital Humanities 2016*, Minneapolis, MN: University of Minnesota Press.

Gordon, Eric, and Paul Mihailidis, eds. (2016) *Civic Media: Design, Technology, Practice*, Cambridge, MA: MIT Press.

Haraway, Donna. (1988) "Situated Knowledges: The Science Question in Feminism and the Privilege of Partial Perspective," *Feminist Studies* 14(3): 575–99.

Hediger, Vinzenz, and Vonderau, Patrick. (2009) "Introduction," in Vinzenz Hediger and Patrick Vonderau (eds.), *Films that Work: Industrial Film and the Productivity of Media*, Amsterdam: Amsterdam University Press, 35–50.

Henke, Nicolaus, Jacques Bughin, Michael Chui, James Manyika, Tamim Saleh, Bill Wiseman, and Guru Sethupathy. (2016) "The Age of Analytics: Competing in a Data-Driven World," Report by McKinsey Global Institute. www.mckinsey.com/business-functions/mckinsey-analytics/our-insights/the-age-of-analytics-competing-in-a-data-driven-world.

Hertz, Garnet. (2012) "Critical Making: Introduction," in Garnet Hertz (ed.) *Critical Making*, Hollywood, CA: Telharmonium Press. http://conceptlab.com/critical making/.

H.RES.51, 113th Cong (2013–2014). Expressing the Sense of the House of Representatives That Adding Art and Design into Federal Programs That Target the Science, Technology, Engineering, and Mathematics (STEM) Fields Encourages Innovation and Economic Growth in the United States. www.congress.gov/bill/113th-congress/house-resolution/51/actions.

Jenkins, Henry. (2004) "Applied Humanism: The Reconstructions Project," *Cinema Journal* 43(2): 91–95.

———. (2009) "What Happened before YouTube," in Jean Burgess and Joshua Green (eds.), *YouTube: Online Video and Participatory Culture*, Cambridge, UK: Polity, 109–25.

Kolbert, Elizabeth. (1985) "The Applied Humanities: A Businesslike Approach," *The New York Times*, April 14. www.nytimes.com/1985/04/14/education/higher-education-the-applied-humanities-a-businesslike-approach.html.

Levine, Lawrence. (1988) *Highbrow/Lowbrow: The Emergence of Cultural Hierarchy in America*, Cambridge, MA: Harvard University Press.

Löwgren, Jonas, and Bo Reimer. (2013) *Collaborative Media: Production, Consumption, and Design Interventions*, Cambridge, MA: MIT Press.

Lupton, Deborah. (2016) *The Quantified Self*, Cambridge, UK: Polity Press.

Maeda, John. (2013) "Artists and Scientists: More Alike than Different." *Scientific American*, July 11. https://blogs.scientificamerican.com/guest-blog/artists-and-scientists-more-alike-than-different/.

———. (2017) "Design in Tech Report." https://designintechreport.wordpress.com/.

Manovich, Lev. (2008–2016) "Software Studies Initiative and Cultural Analytics Lab." http://lab.culturalanalytics.info/.

McPherson, Tara. (2009) "Media Studies and the Digital Humanities," *Cinema Journal* 48(2): 119–23.

Moretti, Franco. (2013) *Distant Reading*, New York: Verso.

National Science Foundation. (2010) Globalization of Science and Engineering Research: A Companion to Science and Engineering Indicators 2010, "Definitions." www.nsf.gov/statistics/nsb1003/definitions.htm.

Neff, Gina, and Dawn Nafus. (2016) *Self-Tracking*, Cambridge: MIT Press.

Orgeron, Devin, Marsha Orgeron, and Dan Streible, eds. (2012) *Learning with the Lights Off: Educational Film in the United States*, New York: Oxford University Press.

Ostherr, Kirsten. (2018) "International Animation Aesthetics at the WHO: *To Your Health* (1956) and the Global Film Corpus," in Christian Bonah, David Cantor, and Anja Laukotter (eds.), *Health Education and Film in the Twentieth Century*, Rochester, NY: University of Rochester Press.

Ramsay, Stephen. (2013) "On Building," in Melissa Terras, Julianne Nyhan, and Edward Vanhoutte (eds.), *Defining Digital Humanities: A Reader*, New York: Ashgate, 243–45.

Ratto, Matt, and Megan Boler, eds. (2014) *DIY Citizenship: Critical Making and Social Media*, Cambridge, MA: MIT Press.

Rheingold, Howard. (2012) *Net Smart: How to Thrive Online*, Cambridge, MA: MIT Press.

Roued-Cunlife, Henriette. (2016) "The Digital Future of Humanities through the Lens of DIY Culture," *Digital Humanities Quarterly* 10(4). www.digitalhumanities.org/dhq/vol/10/4/000277/000277.html.

Rowan, David. (2012) "Reid Hoffman: The Network Philosopher," *Wired Magazine*. www.wired.co.uk/article/reid-hoffman-network-philosopher.

Sayers, Jentery. (2018) "Introduction" in Jentery Sayers (ed.), *The Routledge Companion to Media Studies and Digital Humanities*, New York: Routledge.

———, ed. (2018) *The Routledge Companion to Media Studies and Digital Humanities*, New York: Routledge.

Schulz, Kathryn. (2011) "What Is Distant Reading?" *The New York Times*, June 24. www.nytimes.com/2011/06/26/books/review/the-mechanic-muse-what-is-distant-reading.html.

Shulman, James. (2017) "A Good Job for Humanists," Andrew W. Mellon Foundation, Shared Experiences Blog Post, February 6. https://mellon.org/resources/shared-experiences-blog/good-job-humanists/.

Snickars, Pelle, and Patrick Vonderau, eds. (2010) *The YouTube Reader*, New York: Columbia University Press.

Snow, Charles P. (1959/1961) *The Two Cultures and the Scientific Revolution*, New York: Cambridge University Press.

Steinberg, Erwin R. (1974) "Applied Humanities?" *College English* 35(4): 440–50.

STEM to STEAM Website. (2017) "Mapping STEAM" Congressional STEAM Caucus Interactive Mapping Tool. http://map.stemtosteam.org/.

Williams, Raymond. (1974) *Television: Technology and Cultural Form*, New York: HarperCollins.

2

APPLIED MEDIA STUDIES AND DIGITAL HUMANITIES

Technology, Textuality, Methodology

Lindsay Graham

It is clear that digital humanities (DH) has developed an increasingly strong presence within academic institutions over the past few decades; however, the role of digital humanities within, or in relation to specific disciplinary domains, is perhaps a bit less clear. Nearly a decade ago it had become somewhat ubiquitous to ask 'What is digital humanities?' due in part to the field's almost anti-definitional, all-embracing stance. Marked by a desire for inclusivity, community building, and disciplinary openness, DH conceived itself as an amalgamation, a melding of computation and the humanities. Writing in 2012, Rafael Alvarado insisted that

> there is no definition of Digital Humanities, if by definition we mean a consistent set of theoretical concerns and research methods that might be aligned with a given discipline, whether one of the established fields or an emerging, transdisciplinary one.
>
> *(50)*

Despite this persistent gesture away from a codified and even reified conception of digital humanities, historical and academic practices inevitably shape, at least in part, an emerging field. In its inaugural issue, *Digital Humanities Quarterly* featured a "welcome" from its editors; in this letter to the readers, the editors both acknowledge the transdisciplinarity to which DH aspires as well as those institutional and intellectual practices that potentially complicate such aims:

> Digital Humanities is by its nature a hybrid domain, crossing disciplinary boundaries and also traditional barriers between theory and practice, technological implementation and scholarly reflection. But over time this

field has developed its own orthodoxies, its internal lines of affiliation and collaboration that have become intellectual paths of least resistance.

(Flanders, Piez, and Terras 2007)

To acknowledge the "orthodoxies" and "internal lines of affiliation and collaboration" that have taken shape in the emerging field of DH was not necessarily a reproach on the part of the editors, but rather a means of opening a critical door for fruitful debates on the future of scholarly research and development. And, as the editors had hoped, many deliberations ensued that helped shape and re-shape the contours of the field.

One such dialogue has concerned the boundaries between DH and media studies, and as scholars such as Matthew G. Kirschenbaum note, the fields' converging and diverging interests and lines of critical inquiry, at least in part, result from their differing historical practices and methodologies (6). The concentration of DH, historically speaking, has been on recovering text-based documents and returning those to the public sphere, and in some ways the history of the digital humanities reflects the rise of the personal computer, as many of the personal computers of the 1980s and 1990s lacked the processing power to work with anything other than text. Radio, film, video games, and other audiovisual media that make up the primary objects of analysis for media studies were simply beyond the capacity of the early digital technology accessible to most humanities scholars. It is thus understandable that the self-referential "Humanities Computing" would find a stronghold in English departments in the final decades of the twentieth century.

Indeed, Kirschenbaum reminds us that

> there is a long tradition of text-based data processing that was within the capabilities of even some of the earliest computer systems and that has for decades fed research in fields like stylistics, linguistics, and author attribution studies, all heavily associated with English departments.
>
> *(6)*

Perhaps the most well known of the histories of humanities computing recalls these "earliest computer systems" and the Italian Jesuit priest, Father Roberto Busa, after whom the "Busa Prize," which honors leadership and innovation in DH, is named. Beginning in 1949, Busa embarked on the creation of the *Index Thomisticus*, essentially lemmatizing over 11 million medieval Latin words in the works of Thomas Aquinas and other authors (Hockey 2004). Recognizing the arduousness of his undertaking, he famously approached IBM in the U.S. to seek the technological assistance of a new machine—the computer. With a sizeable team of Italian women, Busa transferred all of the texts to punch cards and proceeded to write a systematic concordance program (Terras and Nyhan 2016). His aim was both to create a tool to conduct text searches and to create printed editions of these concordances (Hockey 2004). Before the completion of his first printed volume in 1974, other scholars with an interest in historic texts began to see the utility of a mechanized concordance creation and attempted their own computer-searchable data systems, though not at the scale of Busa's work.

Advancements in concordance programs prompted a plethora of interest in vocabulary studies and stylistics analyses, and as centers and journals devoted to Humanities Computing were established from the 1960s onwards, methodologies for tagging, searching, and preserving textual data became increasingly familiar to literary scholars (Hockey 2004). With the increased availability of the personal computer and the promises of the internet by the late 1980s and early 1990s, more academics began to perceive the value of the electronic archive as both a supplement and successor to the physical archive. Emerging, for example, from Brown University's English Department, the Women Writers Project (WWP) began to take shape in the 1980s, announcing itself as dedicated to the recovery of texts by early modern women writers. At its inception, the WWP transcribed hundreds of neglected texts, and Oxford University Press printed a 15-volume series based on these transcriptions; the series, *Women Writers in English, 1350–1850*, is still available in print (Women Writers Project 2017). The WWP expanded its efforts in the early 1990s following the improved knowledge of electronic text encoding (ETE) and increased scholarly attention to editorial theory more broadly. ETE, particularly due to the Textual Encoding Initiative's (TEI) expanded guidelines for marking up humanities texts in 1993, not only made visible texts that history had rendered invisible, but proffered a means of preserving the textual data and metadata for future study. In 1999, the WWP electronically published Women Writers Online and eventually published its own guidelines for textual markup as well as the documentation records for the project's encoding practices and theory (Women Writers Project 2017). These freely accessible source sheets and evolving best practices published by the WWP continue to serve as the model for many ongoing, large-scale digital projects.

The digital environment was also attractive to scholars who took a comparative approach to variant texts by a single author. As the constraints of the traditional codex precluded many holistic and functional editions of the work of writers who were known to be avid revisers, digital technologies proffered a solution: the hypertext edition. For that reason, Ed Folsom and Ken Price became interested in humanities computing and electronic archives and editions in order to assuage the difficulties of representing the mutability of Walt Whitman's texts through time and space. Regarding their intent for the creation of the Whitman Archive, Price claims

> the economics of print publication have led previous editors to privilege one edition or another of Whitman's writings—usually the first or last version of *Leaves of Grass*. Our goal is to create a dynamic site that will grow and change over the years.
>
> *(History of the Project 2017)*

The desire to present all of the extant manuscripts and editions of Whitman's work to facilitate scholarly inquiry into both the author's writing and his process necessitated the exploration of the digital environment as a means to adequately represent and engage with multiple source texts in a meaningful way.

Though the overwhelming majority of early DH projects of the 1990s were text-based enterprises, literary scholars whose research focused on multimodal artists were quick to appreciate the benefit of the digital environment as well. Notably, Jerome McGann's study of the Pre-Raphaelite Brotherhood, particularly Dante Gabriel Rossetti, led him to collaborate with John Unsworth, director of The Institute for Advanced Technology in the Humanities (IATH) at the University of Virginia. McGann envisioned a site where Rossetti's literature as well as his paintings and drawings could be simultaneously held, a site where the full weight of the "Pre-Raphaelite" signature could be explored and analyzed. Similarly, William Blake scholars were accustomed to the financial and temporal burden of studying the work of the British writer and engraver. The digital realm, however, offered an alternative site where Blake's various illuminated books could be centralized, searched, and analyzed, thereby obviating the previous need for scholars to travel to numerous physical archives. Thus, like McGann, Morris Eaves, Robert Essick, and Joseph Viscomi sought the guidance of the ITAH to construct a digital repository of the multimedia works of Blake (Archive at a Glance 2017). The creation of The William Blake Archive and The Rossetti Archive garnered much attention for their multimedia, interdisciplinary perspectives, but again, their core interest was in energizing scholarly awareness of canonical literary figures. Further, methodologically speaking, the analytical features of most electronic editions and archives are written in relation to and built upon an original, scanned page of a codex, manuscript, painting, drawing, etc. that is preserved as an image file. In this regard, then, these archives are not unlike traditional scholarly codex editions whereby work is grounded in original textual materials. The digital structure, though, with its focus on numerous source texts as equally weighted points of departure for the user, is one that seemingly eschews an editorial practice that advances or assigns a best or ideal text.

In short, the advent of the personal computer was a boon for textual studies scholars, and the application of statistical and stylistic analysis to text-based research only strengthened the affiliation between computational fields and the traditional humanities, though the utilization of technological tools to advance humanities inquiries privileged a relationship that was somewhat unidirectional in nature. And while the field of DH has benefited from the interest of literary scholars and text-based studies, the demarcation of DH as a field constituted by literary-oriented research inquiries has been at the forefront of debates. Kathleen Fitzpatrick argues that DH "has to do with the work that gets done at the crossroads of digital media and traditional humanistic study" and suggests that this work happens in two distinctive ways (Lopez, Rowland, and

Fitzpatrick 2015). She claims "it's bringing the tools and techniques of digital media to bear on traditional humanistic questions," but she also acknowledges that newer DH investigations and practices are "bringing humanistic modes of inquiry to bear on digital media" (Lopez et al. 2015). More scholars have embraced DH as a melding of or toggling between these modes, though this divergent impulse noted by Fitzpatrick continues to shape much of the critical discourse surrounding the discipline.

Matthew Gold and Lauren Klein's recent volume, *Debates in the Digital Humanities 2016*, foregrounds key issues that confront researchers who work, as Fitzpatrick states, "at the crossroads." These core issues, it seems, have shifted from the first instantiation of *Debates in the Digital Humanities* in 2012. While the earlier focus had been on defining and defending digital methods such as computational analysis, textual markup, and metadata inclusion for text-based research, the contemporary dialogues have taken a more political and intellectually self-reflexive turn. Gold and Klein identify scale and scope as the most pressing topics for DH researchers now, especially the growth of the field itself, the way in which that growth is assessed and measured, and the stakes of digital scholarly research (xi). They ask us to consider, for instance, the openness of digital work to various populations, non-Anglophone languages, or localities with low or no technology. In other words, how high is the barrier to entry in the field and who benefits from its output? These are the questions that are central when considering the problems and potentialities of the digital humanities at present.

As DH continues to expand and redefine its boundaries, scholars such Alan Liu suggest that digital humanists, in order to better understand their field and the role of their scholarship, should

> engage in much fuller conversation with their affiliated or enveloping disciplinary fields (e.g., literary studies, history, writing programs, library studies, etc.), cousin fields (e.g., new media studies), and the wider public about where they fit in, which is to say, how they contribute to a larger, shared agenda expressed in the conjunction and collision of many fields.
>
> *(2013)*

Katherine Hayles' call for an increase in media-specific work is one that resonates with Liu's notion of engaging in broader and richer conversations regarding digital technologies and practices. According to Hayles, DH as a field is, in fact, shifting as scholars "advocate a turn from a primary focus on text encoding, analysis, and searching to multimedia practices that explore the fusion of text-based humanities with film, sound, animation, graphics, and other multimodal practices" (25). Situating the digital humanities in its social and historical context, then, allows us to visualize the intellectual tracks that consider how and where DH diverges from media studies (McPherson 2009), though it gestures also to the present and future commonalities of the disciplines. As applied media practitioners and DH researchers are increasingly pointing to the way we think not only through but with digital tools, it is also becoming more evident that our digital environment is a culturally constructed space. Digital humanities and applied media studies, then, may proffer complementary modes of inquiry and creative making, and in so doing, forge innovative intellectual tracks that fuse text with technology, craft with critique.

References

Alvarado, Rafael C. (2012) "The Digital Humanities Situation," in Matthew Gold (ed.), *Debates in the Digital Humanities*, Minneapolis: U Minnesota P, 50–55.

"Archive at a Glance." (2017) in Morris Eaves, Robert N. Essick, and Joseph Vascomi (eds.), *The William Blake Archive*, www.blakearchive.org/staticpage/archiveataglance?p=archiveataglance.

Flanders, Julia, Wendell Piez, and Melissa Terras. (2007) "Welcome to *Digital Humanities Quarterly*," *Digital Humanities Quarterly* 1(1). www.digitalhumanities.org/dhq/vol/1/1/000007/000007.html.

Gold, Matthew K., and Lauren F. Klein. (2016) "Digital Humanities: The Expanded Field," in Matthew K. Gold and Lauren F. Klein (eds.), *Debates in the Digital Humanities 2016*, Minneapolis: U Minneapolis P, ix–xv.

Hayles, N. Katherine. (2012) *How We Think: Digital Media and Contemporary Technogenesis*, Chicago, IL: U of Chicago P.

"History of the Project." (2017) in *The Walt Whitman Archive*, Ed Folsom and Kenneth M. Price (eds.), http://whitmanarchive.org/about/history.html.

Hockey, Susan. (2004) "The History of Humanities Computing," in Susan Schreibman, Ray Siemens, and John Unsworth (eds.), *A Companion to Digital Humanities*, Oxford: Blackwell, www.digitalhumanities.org/companion/.

Kirschenbaum, Matthew G. (2010) "What Is Digital Humanities and What's It Doing in English Departments?" *ADE Bulletin* (150): 55–61. doi:10.1632/ade.150.55

Liu, Alan. (2013) "Is Digital Humanities a Field? An Answer from the Point of View of Language." *Alan Liu Website*. http://liu.english.ucsb.edu/is-digital-humanities-a-field-an-answer-from-the-point-of-view-of-language/.

Lopez, Andrew, Fred Rowland, and Kathleen Fitzpatrick. (2015) "On Scholarly Communication and the Digital Humanities: An Interview with Kathleen Fitzpatrick," *In the Library with the Lead Pipe*, January. https://doaj.org/article/5d3a0ad130c742e9a408803805fdc438.

McPherson, Tara. (2009) "Introduction: Media Studies and the Digital Humanities," *Cinema Journal* 48(2): 119–29.

Terras, Melissa, and Julianne Nyhan. (2016) "Father Busa's Female Punch Card Operatives," in Matthew K. Gold and Lauren F. Klein (eds.), *Debates in the Digital Humanities 2016*, Minneapolis, MN: U Minnesota P, 60–65.

"Women Writers Project History." (2017) *Women Writers Project*, Boston, MA: Northeastern University. www.wwp.northeastern.edu/about/history/.

PART II
Foundations

3

FOUNDATIONS OF APPLIED MEDIA STUDIES

Kirsten Ostherr, Heidi Rae Cooley, Bo Reimer, Anne Balsamo, Patrick Vonderau, Elizabeth Losh, Eric Hoyt, Tara McPherson, and Jason Farman

Kirsten Ostherr

Each of the collaboratively authored chapters of *Applied Media Studies* was produced through a series of interview-style questions that I, as editor, developed and circulated to the contributors. Through a dialogic process that took place in a deliberately conversational tone, I asked each contributor to answer questions related to a set of themes in the book as a whole, ranging from logistical concerns to methodological and theoretical problems. In this foundational chapter, I asked, "What does applied media studies mean to you? How and why did you start doing applied media studies? In your view, what is the theoretical, historical, and/or political rationale for reimagining humanistic media studies as an applied practice?" In addition to their written responses, contributors created short videos for a web-based companion to the book, hosted on the open-access Scalar platform (http://scalar.usc.edu/works/applied-media-studies/index).

Many of the responses to this chapter's questions noted that their entry into applied media studies came out of a sense of necessity. Several contributors described how the dissemination of their scholarly research through traditional print publications failed to adequately represent the multimediated nature of their work. Others found that their work engaged communities of interest beyond the confines of formal academic institutions, for whom communication of ideas takes place primarily through online participatory modalities. Reflecting the shared sense that transmedia scholarship became necessary to these contributors, certain keywords appeared frequently: interdependency, community, public, collaboration, intervention, and accountability. Perhaps most notably, many of the contributors expressed a sense of surprise and delight at the unexpected circumstance of finding themselves doing applied media studies work. The uncertainty inherent in this experimental approach also serves for many as an engine of discovery.

What Does "Applied Media Studies" Mean to You? How and Why Did You Start Doing "Applied" Media Studies?

Heidi Rae Cooley

Applied media studies works on and for audiences. While it involves active, hands-on engagement with media objects, forms, and/or technologies, communities are its sites of application. Applied media studies draws attention to the varied and complex relations ordinary and not so ordinary media practices establish and sustain, but also revise. It provokes in practitioners, participants, and audiences attentiveness to otherwise familiar or routine transactions that characterize the mobile, connected present in order to cultivate more sophisticated understandings of our media-saturated environments. It does this with an emphasis on the socio-cultural logics that inform how our devices and our use of them position us as members of populations. As a consequence, applied media studies hopes to invite people to reconsider how the status quo—for example, being on grid and locatable—becomes the status quo and the stakes that the status quo serves for reinforcing socially, culturally, and politically established ways of thinking and living.

For me, applied media studies is deeply collaborative, which means it also imagines a different audience than more conventional media studies scholarship. The best examples of applied media studies emerge out of dedicated cross-disciplinary exchange that evolves over time. In my case, my collaboration involves computer scientists and a diverse group of historians (public history, art history, and film and media history), as well as researchers of human–computer interaction and media artists and theorists. Colleagues and students contribute to the development of mobile applications that make visible on location and in real time regularly unacknowledged local histories—histories of slavery and racial politics that have physically shaped the University of South Carolina, Columbia campus. More recently, this collaboration has expanded to include members of the Ward One Organization, an active group of senior citizens who once lived in an area of Columbia, SC, called Ward One, which now is the site of ongoing university expansion. Our approach to applied media studies explicitly attempts to establish an active relationship with people beyond the academy. One might understand this version of applied media studies as a project of public engagement and outreach—offering a counter to more traditionally insular modes of academic production.

I would like to think that applied media studies strives to provoke changes broader than those typical of any academic field of study (e.g., film and media studies, mobilities studies, platform studies, etc.), or traditional venue for academic scholarship and research (e.g., the peer-reviewed article, the monograph), or classroom exercise or assessment (most often, the end of term paper or exam). For me, applied media studies is both scholarly methodology and pedagogy: It is an *ethos*, in the Foucauldian sense of the word, that is, a *way* of scholarship and research and a *way* of teaching. In both cases, applied media studies asks persons to take up the very technologies that recede into the background of habitual everyday use and put them to new, strange, or extreme uses—to test the limits of both our devices and our assumptions about them and their function in the world. For instance, the Ward One application, which is the case study I discuss in this volume, mobilizes geo-location functionality to access a backend database that "pushes" site specific content to the mobile touchscreen device in real time as an interactor approaches a content point, such as the Koger Center, home to the university's School of Music, and formerly the grounds where once stood the birthplace and childhood home of former Ward One resident Mattie Roberson. Here, the navigational interface becomes a portal onto the past. Content "pushed" to screen disrupts any "passive" or rote traversal of the cityscape by inviting interactors to consider the racial politics that underpinned Columbia's federally sanctioned mid-twentieth-century city beautification project that was responsible for the dislocation of a predominantly African American community.

My commitment to applied media studies, which I have historically referred to simply as practice, began very early in my graduate career. Initially a student in the Visual Studies Program at the University of California, Irvine, I quickly discovered that theorizing about visual culture required—or was best served by—some form of practice. At the time, I was interested in mobile screen devices and the ways such devices shaped engagements with and traversals of one's surroundings. My first object of study was the camcorder with pop-out LCD screen. To intervene in various discussions—both scholarly and not—about the implications of the LCD screen extension to the consumer video recorder, I purchased such a device; and I documented, but also mimicked, the practices I observed. I continued this approach when I shifted my intellectual focus to the camera phone. Referring to myself as a participant-observer—someone who practices what, at the same time, she watches—I considered, in very concrete but theoretically informed ways, what it meant to image on-the-go and post to image sharing sites such as Text America. What new ways of experiencing and documenting and sharing were evolving? And what might this mean? Why might this matter—to whom? The Ward One project continues this emphasis but in the context of collaboration, where the challenges and virtues of a team allow scholarly practice to be "applied" in ways that have the potential to invite a public, by means of the navigational interface, to reconsider how the past affects the contemporary landscape.

Bo Reimer

For me applied media studies means combining theoretical, critical, and empirical work with design oriented, experimental work focusing on the not-yet existing.

It means setting up situations in which researchers from different disciplines and with different competencies together with other, non-academic actors intervene in public sphere activities, or in any other way try to make a difference in regard to questions with social, political, economic, or cultural relevance.

It is a perspective dealing heavily with notions of change. Instead of only reacting upon and analyzing social, political, or cultural situations, researchers intervene in these situations, starting processes that may not have started otherwise. In other words, applied media studies is about leaving the objective researcher role behind, instead taking a stand in important societal questions, as well as assuming responsibility for one's actions. Important keywords are collaboration, publicness, intervention, and accountability.

Without using the term explicitly, I started doing applied media studies more than fifteen years ago, having gotten increasingly tired of the traditional social science/humanities media researcher position.

The traditional position is one where you theorize, analyze, interpret, and produce conclusions, telling others: "This is how it looks", or "This is how it is." Additionally, you may add normative statements, based on the analysis: "It should not be like this" ("the working situation for journalists is terrible", "the quality of the media output is terrible", etc.). And maybe even creative, constructive statements: "This is how it should look, and if you do this...."

The latter stance makes explicit an interest in how things are, as well as an interest in the possibilities for change. In that sense, it is a way of taking part, taking a stance. But it is a way of doing it from a distance, from the outside. It is also based on reacting to things already there or already happening.

Such a stance is valuable, of course. But it could—and should be discussed: Is it possible to stay on the outside? And is it really the best position to take?

Staying on the outside could be gratifying for a media researcher, for instance, telling news media producers how badly and naively they carry out their work. That was also my feeling initially. However, increasingly I felt dissatisfied with such a position, instead thinking that it would be more constructive and rewarding to actually take part in work that in itself had an impact on the surrounding society.

This turnaround coincided with me participating in the creation of a new transdisciplinary research environment, the School of Arts and Communication at Malmö University, Malmö, Sweden. This environment was built on the idea of making it possible for academics coming from media studies, technology, design, and the arts to work together, both in relation to education and research. These are fields with many interests in common, but for disciplinary reasons they are normally located in different departments, or even in different faculties. The idea was thus to break with these taken for granted divisions, creating a space for new ways of working together. This has now been my home since 1998, and where we together have created what we initially in a manifest called "A digital Bauhaus."

Let me finish by quoting from our manifesto (written in the fall of 1998):

> What is needed is humanistic and user oriented education and research that will develop both a critical stance to information and communication technology, and at the same time competence to design, compose, and tell stories using the new mediating technologies.
>
> What is needed are meetings between:
>
> constructive knowledge and competence related to interactive and communicative possibilities and constraints when using the new mediating technologies,
>
> aesthetic knowledge and competence from fields such as television, theatre, film, music, literature, architecture, art and design, and
>
> analytical-critical knowledge and competence from philosophy, social science, and not least cultural and media studies.

Anne Balsamo

Applied media studies addresses the horizon of possibility where the insights of media study are used to build (new, novel) media experiences that are designed, from the beginning, to be transformative of current cultural arrangements; to be more liberatory, to be more respectful of histories and publics who are lost, voiceless, and subjugated. This approach manifests a social and cultural agenda to intervene in the conditions of possibility that define what can/cannot be addressed through the development of new technologies. In my view, applied media studies also involves the creation of technology prototypes, either devices, applications, systems or hybrid efforts that creatively reconfigure the scene of communication, and culture (more broadly).

Patrick Vonderau

In my understanding, the term "applied media studies" relates to the current flurry of disciplinary cross-border initiatives rather than to an emerging discipline. It points to new networked relations between media studies and other fields where humanist skills and the humanities approach of problematizing, rather than solving life-world issues is seen to produce useful knowledge. It has little to do with what would be its direct German translation, "Angewandte Medienwissenschaften"—a term reserved for emphasizing the practice-orientation of media management and other study programs. It's just a different form of knowledge production, critically theorizing social, technical, industrial, and other aspects of today's mediascape "from the ground up," often in ways that aim to transform this mediascape, in order to identify alternatives where others see none, and to provide different analytical tools that may be helpful for dissecting what so often is just taken for granted.

For me, there is a difference between applied media studies and the digital humanities (DH). As I see it, DH in many ways continues (and amplifies) traditional humanities practices such as collection-building, creating statistical items, scaling data, measuring, localizing, contextualizing, scholarly networking, and the like—common practices even in analog historical or analytical media work. Since the 1980s, we also have seen many great initiatives to develop computation-based forms of media scholarship, in the form of notational systems, databases, or web-based communication among researchers. Yet while certainly related, "applied media studies" goes beyond that in creating forms of knowledge that are essentially networked between various disciplines and practices. It is as if media studies had undergone not just one "turn," but actually three—a turn to the social and the everyday, a turn to method, and a turn to digital media technology.

My interest in this kind of disciplinary borderwalk goes back to 2001 when I was an Assistant Professor at the University of the Arts in Berlin. I joined this university that offers study programs in architecture, art, music, design, experimental filmmaking, and media/communication as a rather traditionally working media and cultural historian, although I had a few years of experience in analogue filmmaking, especially in animation. All those distinctions didn't matter that much anyway at this department, called Institute of Time-Based Media, a collaborative space allowing for exchange between graphic designers, computer artists, underground filmmakers, and scholars like myself. Since this was my very first job, I thought universities just were like this. We had people such as Alan Kay or Nicholas Negroponte visiting for design classes and design students trying to write academic papers. These constant processes of translation between various communities of practice were highly instructive.

In 2001, professor Carlos Bustamante (a filmmaker and Early Cinema historian) and I applied for a small project grant for developing a new application in the field of "rich media content." Someone at faculty level had persuaded Deutsche Bank to donate money for experimental work on how to integrate moving images into Internet-based, interactive systems. This was still a few years before YouTube, and the models we oriented ourselves to were Manovich's work on databases, Steve Mamber's early web projects (http://mamber.filmtv.ucla.edu) and interactive CD-ROMs such as Lauren Rabinovitz' *The Rebecca Project* (1995) or Yuri Tsivian's *Immaterial Bodies: A Cultural Anatomy of Early Russian Films* (1999). The project we got funding for was entitled "e-analysis" (as a riposte to "e-business," then the catchphrase of the day) and aimed to engage film students in using Macromedia Director for building interactive learning environments. In many ways, this project was the direct opposite to today's "digital tools" developed in DH-based media history projects: qualitative rather than quantitative in its approach, and insisting that film analysis should be geared towards what is uniquely significant rather than statistically relevant.

A few years later, Pelle Snickars and I edited *The YouTube Reader* (2009). Academic books often come too late and are boring, we felt. So we wondered if there were ways to overcome standard procedures of academic publishing. YouTube was not just a timely subject, but also one whose dynamic and deeply fascinating character we wanted to translate into scholarly debates and artistic engagement. So we decided to manage the entire book production— including design and printing—ourselves. Curator Giovanna Fossati developed a complementary web exhibit, together with a group of Berlin-based Flash programmers (youtubereader.com).

I mention these two instances because they became formative experiences, testifying to the freedom that is possible—the freedom of creating a space for your work that goes beyond the traditional institutional or disciplinary framework of academic scholarship. And also in order to underline the probing, uncertain, disconnected, and exploratory character of this kind of work. Labels such as "applied media studies" or "digital humanities" emphasize strategy and disciplinary cohesion while it was precisely the lack of the latter that made such work fun and insightful.

I had a third formative experience as a team member in the Connected Viewing Initiative in 2011–2012, a research project led by Jennifer Holt at the Carsey Wolf Center, University of California at Santa Barbara. The project initiated a much-needed critical conversation between industry and academia. To my knowledge, this "teaching back" the media industries was a first in history. Sitting in a Warner Bros. boardroom and hearing those executives explain and defend their corporate culture was enormously instructive. Again, a border-crossing of sorts that turned out to be a unique learning experience.

Elizabeth Losh

My life in applied media studies began when I graduated from Harvard in 1987 and took my first job as "Educational Director" of a delinquency prevention center run by a non-profit foundation that contracted with the California Youth Authority. Despite my impressive title, essentially my position meant that I ran a computer lab. Our clients were mostly young people of color targeted for diversion programs by law enforcement, social service agencies, or public schools. As an idealistic twenty-one-year-old who wanted to experiment with different ways to improve my students' digital literacy, I probably didn't always understand—or respect—the existing knowledge practices in those communities and wasn't adequately self-reflective about my techno-missionary zeal.

I wrote about my experiences using PEN (the Public Electronic Network) at the youth center in the beginning of my first book *Virtualpolitik*. There I described how in retrospect I appreciated the politics of how participants appropriated and subverted the technologies that I helped make available to them, and I considered what their actions implied about Silicon Valley's ideologies of direct democracy, transparency, and neutrality. This first job also probably helped me on my later administrative path as a writing program administrator, where I was thinking about improving college composition curricula, which lamentably often ignored the forms of writing that were most important to students—namely everyday composing for digital networks. This brought me to digital rhetoric as a field invested in the pragmatic and political stakes of computer-mediated communication.

In general I think applied media studies often performs what Leigh Star and Geoffrey Bowker have called an "infrastructural inversion" by investigating questions about interdependency and mess. In its practices, applied media studies deploys and appropriates many forms of action research, restorative justice, urban pedagogy, and policy intervention, while recognizing the mediated character of these different tactical engagements. After all, we can probably never get outside of techno-social environments, matters of the apparatus, and the interfaces and platforms that facilitate interoperability, so applied media studies does not claim to be more authentic than its more established counterpart. For example, in redirecting media studies from a focus on invention to a focus on delivery, the work of Lisa Parks and Nicole Starosielski on media infrastructures requires not only attention to the geography of cell phone towers and Internet cables around the world but also to the perils of uncritical philanthropy, tourism, and the global development paradigm.

In the case of FemTechNet, which was founded by Anne Balsamo and Alexandra Juhasz in 2013, this means teaching about, through, and with computational media—as well as many pre-digital technologies—and giving students a role in debating the affordances and constraints of the design of digital environments as communicative spaces. In the context of FemTechNet classes, students have critiqued and created content for learning management systems, Wikipedia pages, online video channels, teleconferencing platforms, interfaces for writing programming code, Google maps, mobile phone applications targeting youth that promise ephemeral posting, curated photo sites for performing self-representation and applying filters to images, and even workflows for captioning class conversations to promote official access for disabled students. By using the structure of the DOCC (Distributed Open Collaborative Course) rather than the MOOC (Massive Open Online Course), students may be more able to challenge the social contract governing expectations about participation, call out prevalent online misogyny and racism, and bear witness to the harm done by digital microaggressions.

In doing so, FemTechNet hopes to undermine dominant narratives about media technologies and show the applicability of feminist theory to many domains of cultural representation and technological responsibility. Theoretical touchstones include the following assertions:

Technology is material (although it is often presented as virtual).

Technology involves embodiment (although it is often presented as disembodied).

Technology solicits affect (although it is often presented as highly rational).

Technology requires labor (although it is often presented as labor-saving).

Technology is situated in particular contexts (although it is often presented as universal).

Technology promotes particular values (although it is often presented as neutral).

Technology assumes tacit knowledge practices (although it is often presented as transparent).

The challenges that face FemTechNet are familiar ones from the vantage point of acknowledging the importance of care and repair. Reproducing conditions of precarity and exploitation are always a risk of feminist work, and FemTechNet is no exception, particularly when distributed networks depend on unlimited stores of immaterial labor and the availability of access to multidirectional computational media in the home, across time zones, and in ways that can colonize friendship and hospitality with work's intimacy. Sustainability and equity in co-facilitation continue to be issues, particularly in the area of funding and supporting the critical roles of community managers, who are often discounted in media organizations and affinity groups.

The rest of my history with applied media studies has involved other networks that intersect with FemTechNet, such as the Selfie Researchers Network (which organized the Selfie Course) and the Digital Media and Learning Research Hub (which organized Connected Courses). My own research on hashtag activism and witness journalism has also involved collaborating, co-authoring, and co-teaching with activists working on citizen media issues, such as Jasmeen Patheja of Blank Noise and Sam Gregory of Witness.org.

Eric Hoyt

Service is the first word that comes to mind when I think of applied media studies.

Applied media studies is about channeling our curiosities and expertise into various forms of expression that serve communities inside and outside the academy. As such, applied media studies requires breaking down the distinction between "research" and "service"—two different categories on a professor's CV, with service typically valued less for hiring and promotion than research. Applied media studies projects have a service mission, and that mission requires continued public engagement long after any initial moment of publication or online launch.

I started doing applied media studies in 2011 when David Pierce, the founder of the Media History Digital Library (MHDL), gave me the opportunity to get in on the ground floor of that project. The goal was to digitize out-of-copyright fan magazines and trade papers related to film and broadcasting history for broad public access. David had a network of relationships with collectors, institutions, and sponsors that made this project possible. What I had was a car and a computer in Los Angeles.

I spent the first few months driving to a collector's home in Southern California, boxing up historic magazines, delivering them to a scanning center, and then unboxing and returning the publications to the collector's shelves. I then got involved in the digital presentation side, designing the MHDL's website and search platform, Lantern (discussed in depth in my later chapter). The common threads of the physical and digital work that first year were being good custodians of the historical artifacts and trying to offer something valuable to the partners who supported us (complete packing, delivery, and unpacking service) and the online audience we hoped to engage.

I will close by mentioning the second word that comes to mind when I think of applied media studies: collaboration. I don't know of any impactful applied media studies projects that have been built and sustained single-handedly. And besides, these projects almost always require a great deal of learning, working, laughing, and crying. Why would you want to do those things alone?

Tara McPherson

My interest in "applied media studies" unfolded during graduate school, along two interrelated paths. First, I pursued a Ph.D. to further my interests in feminist film theory. I chose the University of Wisconsin-Milwaukee because it was a vibrant center for the field at the time. While my program was in an English Department, UW-M also offered a strong experimental film department through the art school. I was lucky enough to take a course in feminist film team-taught by the theorist Patricia Mellencamp and the video artist Cecelia Condit. The class brought together students from different departments and encouraged collaboration and risk taking. I had never made a film before, and I found the experience exhilarating. I had already begun to analyze and write about film, but the opportunity to experience media production at the material level deeply impacted how I would come to understand my analytical work as well. Working in media production—from cinematography to editing—can lead a scholar to a deeper engagement with both the material form and the audio-visual nature of film and video. I don't think it's an exaggeration to say that the experience rewired me. Throughout that semester, the quality of my dreams shifted; they became increasingly cinematic and deeply saturated. They utilized a filmic language.

I firmly believe that the work of hands-on making, be it on or offline, produces a different type of knowing than the work of writing. For me, these different ways of knowing are equally fruitful and ever entangled. They shape one another in important ways. An "applied media studies" that short shrifts history and theory would fall short of what I see as valuable in such an approach.

The context and content of the class I took long ago were as important as the making itself. Theory and practice were deeply intertwined throughout the term, existing in rich feedback loops with one another, particularly through the lens of feminism. Feminist film studies emerged from an environment that was political and that held theory and practice in productive tension. Practitioners and theorists were in deep conversation about what feminist film could do in the world. Laura Mulvey, author of a landmark essay in feminist film theory, was herself a filmmaker. I suspect her attention to the camera's gaze was in no small way born of her experiences behind a camera. Our class was diving deeply into feminist debates about the status of representation and spectatorship and about the possibility for a feminist aesthetic. Our thinking about these issues shaped how we made media, but our media making also remade our theory. As a professor, I have spent most of my career working in a cinema school. While not all of my colleagues cross the theory/practice divide, some do, and I continue to value such hybridity. We have recently launched a new Ph.D. program that allows students to "make" dissertations that meld theory and practice. I find their work to be very promising and exciting.

The example above hints at another way in which "application" and "practice" matter to me. "Practice" becomes another way of not only understanding *the matter of media*, i.e., its material form. It also becomes a way of *making media matter* in the world. Throughout graduate school, I studied feminist film theory and practice while also engaging in feminist activism on issues of reproductive rights, labor politics, and community media production. At the time, feminists were querying theory's role and impact in the movement as feminism increasingly took root within the academy. My activist work, much like my experiments in feminist media making, helped me take up these questions in an embodied, if often tense and imperfect way. Of course, I believe that theory is political and has effects in the world. I am not suggesting that all feminists should "apply" their theory in some form of material praxis. But I do know that applied media studies can enact theory *differently*, sometimes exceeding the terrains of theory in delightful and important ways. This excess can itself become a powerful fuel for theoretical inquiry even if these two ways of knowing are not entirely commensurable. So, I'd say my interests are in an applied media studies that melds theory and practice in an overtly political manner that reaches beyond the confines of the university. Surely this matters immensely in the world today.

Jason Farman

There are certain questions that can only be explored through hands-on engagement with material artifacts. Ways of knowing require embodied engagement in order to fully analyze the range of implications being studied. So, applied media studies is more than simply learning practical tools for making things. Instead, it is about exploring ideas, questions, and larger cultural implications through gaining access to these issues in an embodied mode of analysis. Much of what I study centers on technology in everyday life, and it is in the everyday where many of the larger cultural implications of our media get buried. A lot of the power and specifics of how our media function resides in their ability to exist in the everyday and tap into our practices of common sense. The "common sense" I refer to here draws on Antonio Gramsci's (2000) exploration of the ways that hegemony exerts control not through coercion but through incorporation into common sense. So for me, applied media studies is about exposing the everyday assumptions that we have about our media. By opening up our machines (literally and figuratively), by understanding how they work, by building and making, and by exposing the various levels of infrastructure required to make our media function, we can begin to pull back the veil of the everyday and expose the power of "common sense."

In Your View, What Is the Theoretical, Historical, and/or Political Rationale for Reimagining Humanistic Media Studies As an Applied Practice?

Heidi Rae Cooley

Through application, media studies becomes ethical in the Foucauldian sense of the term. It is an *ethos*: a way of study that acknowledges and accounts for the very material impacts that such study might, does, and should produce. It is to understand and establish one's study as an evolving relationship between one's self and one's disciplinary field/s, one's objects of study, and the world—people, histories, landscapes—beyond study as such. It involves constant negotiation and responsiveness because the application of study through practice is lived and experienced and, in the best versions, shared—and therefore, deeply personal. The result is a process of study that is responsible to a broader community of collaborators and/or participants. It remains important to me that professional scholars consider my work accurate, well informed, and insightful, but in the end, it matters more that larger communities of people come to understand and engage—in more considered ways—the world they inhabit.

Bo Reimer

I regard reimagining humanistic media studies as an applied practice as a primary political project; a democracy project.

Our contemporary media landscape is a collaborative media landscape; a landscape characterized by the possibilities for citizens to not only consume media but also take part in the design of the media infrastructure and to produce their own content. This, however, is carried out within a heavily structured framework. Citizens have the freedom to work and to publish on the Internet. But they are not in control: they are at the mercy of private Internet companies.

The questions for the media researcher are then: How can we best make sense of this? Can we do something about it?

Here is where I believe a practice based approach becomes crucial. Even though the collaborative media landscape is dominated by a relatively low number of private companies, there is an ongoing struggle over how this landscape should look, and who the determining actors should be; the outcome of this struggle is not predetermined. As technologies for media creation become easier to use, not only citizens and professional media producers but also media researchers are able to take part in constructive experiments "in the wild," adding their specific competencies to the mix. Thus, by taking part in coalitions with other actors, integrating intervention, analysis, and criticism, a practice based media studies can make a difference.

Anne Balsamo

Protocols of technology development and knowledge creation are undergoing transformation through the use of digital media, the development of new genres of digital scholarship, as well as the creation of new partnerships between the university and multiple publics as collaborators and audiences. Humanistic engagement with applied media practices often focus on the development of new meaning-making applications, tools, and devices. Because they focus on new forms of meaning-making and new protocols for the circulation of meaning, these projects are implicated in the development of new literacies. The development of new literacies is, fundamentally, a political project because it is the foundation for future world-making.

Patrick Vonderau

I am not entirely sure if there is indeed one rationale, given the differences between our experiences and academic cultures. Insisting on a rationale could also prematurely channel or stop a process that is very much open-ended. I think, however, that we might want to reimagine humanistic media studies as a form of knowledge production that invokes new "trading zones," to use a term coined by Peter Galison. Galison insists on asking how, at a given linguistic border or "interstitial zone" between various academic subcultures, new terminology evolves in the coordination of various values and meanings. Although Galison was mainly interested in the history of physics and the production of laboratory knowledge, his insights matter for scholars moving between disciplines as much as between academia and other spheres of social life. For

Galison, these encounters between subcultures produce a form of knowledge that cannot be explained in terms of great ruptures, confrontations, or paradigm shifts. Rather, knowledge emerges over many subsequent encounters and gradual coordination between various domains of practice on a local level.

As the contributions to this book demonstrate, there is a huge variety in how media studies is transformed locally through encounters with society, technology, and the like. In order to keep this exchange open-ended and alive, we may want to acknowledge the local and historical character of these initiatives. In my view, there is no lack of new disciplinary subfields and proclaimed paradigm shifts in media studies. Put differently, instead of focusing on emerging paradigms, we may want to focus on keeping the border traffic alive. If "applied media studies" will last, it is precisely because of a generational and institutional change that allowed it to open up for such border traffic in the first place.

Elizabeth Losh

Applied disciplines often are perceived of as having less prestige than their non-applied counterparts. For example, a department of applied linguistics may be derided as a service unit for non-native speakers of English, despite its critical importance to its constituents. Before computer science was seen as a lucrative unit in the university, students pursuing software programming might have majored in the less glamorous field of applied mathematics, which was seen as an academic stepchild, particularly in the Ivy League.

Surrendering the status (and snobbery) associated with media high theory to embrace applied media studies better aligns the scholarly labor of its practitioners with their stated political ideals and more accurately situates media studies with other forms of critical making historically. After all, contemporary media studies must be engaged with living communities as sites of knowledge production to remain relevant and be committed to inequality research, which requires fieldwork, ethnographic investment, deep forensics, and often making a tangible case for compensation rather than armchair philosophizing.

This reorientation will also produce better theory, as it will better practice. To explicate more rigorously the material, embodied, affective, situated, and labor-intensive conditions of media and mediation requires interrogating abstractions that are accepted as received ideas.

Eric Hoyt

For me, the best articulation for why "applied" research matters comes from over a century ago. In 1904, the President of the University of Wisconsin, where I work, said the following: "I shall never be content until the beneficent influence of the University reaches every home in the state." President Van Hise's philosophy became known as "The Wisconsin Idea."

The Wisconsin Idea has been interpreted to signify an outreach mission for research and teaching that begins with the state of Wisconsin and extends nationally, even globally. It's an aspirational and progressive mandate, calling on scholars to translate their research into products, services, and conversations that change people's lives for the better. Scientists pursuing treatments and cures for diseases are pursuing the Wisconsin Idea, certainly. Humanists have a role to play too. Many of us curate public exhibits (some physical, others virtual), give talks outside of academic settings, and engage in conversations about media, society, and citizenship. But we still have a long way to go. All faculty members across the disciplines, myself included, could be spending more of our time engaged in work that benefits a public beyond the ivory tower.

I had been working on the Media History Digital Library and Lantern for several months before I was offered a job at the University of Wisconsin-Madison. The potential for these projects to provide broad, free access to historical documents and to reach tens of thousands of people was a big part of why I was passionate about working on them. When a senior faculty member told me about the Wisconsin Idea, I remember thinking, "Yep, I belong here."

Over the last few years, it's become clear to me that the Wisconsin Idea is not something to be taken for granted. In 2015, Wisconsin's governor attempted to remove the Wisconsin Idea from the state university system's educational mission. In place of this public service mandate, the governor wanted to more narrowly define the University of Wisconsin System's mission as being one to "meet the state's workforce needs." Faculty, students, journalists, and state citizens protested this rewriting of our educational mission. The Wisconsin Idea was officially reinstated, even though severe budget cuts and a weakening of tenure protections were implemented at roughly the same time.

I bring up this last story, in part, because I've grown frustrated to hear some colleagues casually dismiss digital humanities initiatives as being cogs in a twenty-first century neoliberal takeover of American universities. I can only speak for myself, but my digital work is rooted in a vision of public service articulated in 1904, and I plan to keep fighting for that vision and attempting to enact it in my work.

Tara McPherson

The history of feminist media studies in the U.S. and the U.K. suggests that the field had strong ties to practice at the moment of its formation as an academic discipline. In the 1970s as the field began to take shape in the west, feminists were also responding to a shifting technological field. They were at once confronting the popular spread of television and discussing its limits while also taking up newer, more portable production equipment to produce images of their own. They understood that they needed to create their own possibilities for distribution and exhibition for feminist media and also to work within mainstream media. They engaged the technologies of their moment, and these hands-on engagements shaped the ways in which they imagined and wrote theory. As the field has settled into the academy over the past few decades, it has not always been easy to sustain a dialectical relationship of theory to practice. While some feminist media scholars continue to cross over into making, many do not.

The rise of digital media studies offers an opportunity for media scholars to reengage practice on a number of levels. This is especially true for scholars concerned with issues of gender, race, sexuality, and social justice. I think there is also a real urgency here, much as in the 1970s. The wide diffusion of digital technologies in the past two decades has brought rich media making tools into the palms of our hands while also raising pressing questions about surveillance, privacy, access, justice, and labor. Women are again faced with pronounced struggle for reproductive and other human rights. Public conversations about sexual and racial violence saturate our media. If feminists in the 1970s understood the importance of "seizing the means of production," we might do the same today as feminist media scholars. By taking up digital technologies and striving to bend them to our desires, we will both better understand the machines that powerfully shape everyday life today and also perhaps remake those machines to better serve our needs. Reimagining humanistic media studies as an applied practice helps capture this sense of urgency while also reactivating the early history of feminist media studies.

Jason Farman

My approach to applied media studies draws on a range of material from phenomenology to theories of critical making and cultural studies approaches to media. My work here is founded on an idea that I first encountered through cultural geographer Raymond B. Craib (2000), who described the power of maps to exert influence and to maintain specific ideologies. In discussing the power that maps (as a medium) can exert, he says, "No other image has enjoyed such prestige of neutrality and objectivity. The most oppressive and dangerous of all cultural artifacts may be the ones so naturalized and presumably commonsensical as to avoid critique" (8). Here, Craib builds on Gramsci's notion of hegemony. Power is often exerted by the things that we do not question, and many media in our everyday lives fit this mode of engagement. In my own work studying the body and mobile technologies, I draw a connection between this approach and some of the work of phenomenologists. Here I'm thinking specifically of philosopher Martin Heidegger's notions of ready-to-hand and present-at-hand, in which a tool that is incorporated into the body goes unnoticed (or is unremarkable) (Harman 2002). It is often when the tool breaks that we begin to question and critique its incorporation into our bodies and our everyday lives.

Applied media studies is about this work of moving from ready-to-hand to present-at-hand, or what Jay David Bolter and Richard Grusin (1999) would call the move from immediacy to hypermediacy. I trace this kind of work in media back through many theorists' approaches to the study of media. For example, in my doctoral research in the field of performance studies, the work of Bertolt Brecht started me down this intellectual pathway. Brecht's (1957) political approach to creating and analyzing theater as a medium aimed to remove its modes of immediacy—the defining features of realism at the time—to redefine theater as a medium that audiences engaged with. This extended into the performance work of people like Augusto Boal (1985) and his approach to the Theater of the Oppressed, in which the audience members were the creators of the theater. This is applied media studies in the truest sense, when the work of creating a medium allows you to see the political and cultural implications of that medium. Making is a mode of thinking. Making becomes a way of connecting our bodies to the medium, of uncovering the assumptions of those media, and of imagining new ways of using the media.

References

Boal, Augusto. (1985) *Theatre of the Oppressed*, New York: Theatre Communications Group.

Bolter, Jay David, and Richard Grusin. (1999) *Remediation*, Cambridge, MA: The MIT Press.

Brecht, Bertolt. (1957) John Willett (ed. and trans.), *Brecht on Theatre: The Development of an Aesthetic*, New York: Hill and Wang.

Craib, Raymond B. (2000) "Cartography and Power in the Conquest and Creation of New Spain," *Latin American Research Review* 35(1) (2000): 7–36.

Gramsci, Antonio. (2000) David Forgacs (ed.), *The Antonio Gramsci Reader: Selected Writings 1916–1935*, New York: New York University Press, 2000.

Harman, Graham. (2002) *Tool-Being: Heidegger and the Metaphysics of Objects*, Chicago, IL: Open Court.

4

FROM *VECTORS* TO SCALAR

A Brief Primer for Applied Media Studies

Tara McPherson

> Designing always offers an opportunity to do something that has not been done before and to create something unique and untried. Designing provides participants with the tools to exercise their technological imaginations in the creation of our collective futures (7).
>
> —*Anne Balsamo (2010)*

A few years ago while away at a conference, I received an urgent text message from Curtis Fletcher, the project manager for Scalar, an online publishing platform for which I am a principal investigator. He had received an angry message from a group of young women who had stumbled upon a Scalar project describing their participation in a pro-anorexia online community. The piece had been completed without their consent or knowledge and included information that they felt violated their privacy. They asked that we immediately take down the site. Scalar is a publicly available, free online resource with over fifteen thousand users, and this project had been created by an author our team did not know. We found ourselves needing to adjudicate between the young women's claims to privacy and the author's use of our platform. As I detail below, this particular case was resolved rather easily, but it raised important issues for our team about software development and online publishing that we had not explicitly planned for, nor addressed before. It is one of several lessons learned from over fifteen years working in applied media that I will describe in this chapter.

The platform Scalar went live over five years ago, building on the work of the team behind the multimedia journal *Vectors*, alongside many other collaborators. The core team working on both projects includes Steve Anderson,

Craig Dietrich, Erik Loyer, and me. Phil Ethington, Curtis Fletcher, and Lucas Miller have been integral additions to Scalar's recent development, while Raegan Kelly was central to the early years of *Vectors* and to the initial planning that led to Scalar.[1] Both projects are ongoing experiments in open access publishing, emerging screen languages, and academic software development. They are motivated as well by a deep engagement with progressive scholarship within the interpretative and theoretical humanities, particularly in such fields as gender and sexuality studies, media studies, African American and ethnic studies, post-structuralism, and so on. From our early experiments at *Vectors* to our current support for diverse forms of online authoring, our efforts have centered on bridging theory and practice. In *Where the Action Is: The Foundations of Embodied Interaction*, Paul Dourish (2001) observed that theory and practice can be hard to hold together because "[t]he goals and criteria for theoretical examinations are quite different from those for design exercises" (155). He is arguing for designers in fields like human-computer interaction to create a more explicit relationship to theory because "theory makes design real" (158). In our ongoing work with humanities scholars, we have found that emerging forms of practice can also enrich theory-driven disciplines when the relationship between the two modes of working is deliberately pursued. These endeavors have not been without friction of many kinds. In what follows, I detail some of the tensions that have emerged across years of our work, framed as a five-part primer for undertaking applied media studies.

Tip One: Scaffold Collaborations across Difference

Vectors began as an investigation into interactive publishing around 2003, at a time when online scholarly publishing had not ventured far beyond text with pictures.[2] As we conceptualized the journal, Steve Anderson and I aimed to support projects that were interactive and multimodal, pieces that could not exist on the printed page and that were fairly experimental in design. We were after something quite unlike typical journal articles and were mining the boundaries between humanities scholarship and the arts. We soon realized that very few humanities scholars would be able to create their own projects for the journal, so we looked for models of collaborative design that we could adapt for the projects we would develop in house. We were intrigued by the Labyrinth Project at USC, an initiative spearheaded by our colleague Marsha Kinder, that brought established literary or film artists together with an interactive design team to create new digital art works.[3] We soon settled on a fellowship process that would allow humanities scholars to work closely with a small design and editorial team to create digital scholarship. Each core team consisted of the scholar, an editor (Steve or I), and a designer (typically Raegan Kelly or Erik Loyer); Craig Dietrich worked closely with the designers and editors throughout the process and also built the journal's backend. These collaborations spanned several months, beginning with a one-week summer 'boot camp' for face-to-face design meetings and continuing with ongoing virtual interactions. These collaborations were the heart of our early process for *Vectors*; they were exhilarating and often also challenging, particularly because they were bringing together artists, technologists, and scholars with very different competencies and methods of working. We soon developed strategies for supporting this process.

First, we learned the value of shaping expectations early in our interactions with scholars. Our process was highly iterative, allowing for a fair amount of exchange and give and take between editor, designer, and scholar. There was no central formula at play. But even this very iterative process needed scaffolding, and we were gradually able to refine our working methods. From the outset, it was important to underscore that our team wasn't working *for* the scholar but *with* the scholar. We made sure that scholars knew that they would be helping to build their projects alongside our team, interacting with our software and helping to shape the ways in which their scholarship would manifest in digital form. Some scholarly technology centers implement a scholar's work on his or her behalf. This was not the goal for our team. Our process was more integrative and not based on a "design for hire" or service model. To facilitate this process, we created a framework for use in early design meetings. We asked scholars to consider the following before their first sessions:

1. Conceptual issues: key arguments from their scholarship; is the project "born digital" or translated from print; relevant fields of inquiry
2. Aesthetic direction: thoughts on overall look and feel; possible models that the aesthetic design might draw from; affective tone desired
3. Interactivity and information design: desired user experience; how might the project unfold; is a database needed
4. Intersections of form and content: how do 1–3 interconnect; what do you envision publishing in *Vectors*
5. Function of the project: is it archival, experiential, argumentative, explorational, spatial, immersive, etc.?
6. Metadata and keywords
7. Existing assets: clips, images, articles, maps, sound files, etc.
8. Logistical issues: rough production schedule, workflow, modes of collaboration, and team delegation
9. Technical requirements: software and hardware, database needs, hosting location, formats and compatibility, existing development resources

While these questions are obviously broad and overlapping, they helped give our initial meetings direction without forcing each project into a specific template or rigid structure. They also began the important process of sculpting a shared language across the team. In her wide-reaching research on interdisciplinarity, Julie Klein (1996) observes that truly integrative research "is neither routine nor formulaic. It requires active triangulation of depth, breadth, and synthesis" (212). Our design process was oriented toward a similar model, iteratively bringing together deep disciplinary knowledge, multiple perspectives, and varied skill sets toward an outcome that uniquely synthesized and expressed diverse ways of knowing. Discussion of these parameters created a space from which a mutual vocabulary might be built up for each project, allowing the different team members a chance to discuss ideas, question assumptions, and find common ground. The editors took a more active role at the beginning of the process, helping to translate across domains between scholar and designer. After the work on each project took root, such translation was only needed at key junctures. In *Crossing Boundaries*, Klein notes that "difference, tension, and conflict emerge as important parts of the integrative process. They are not barriers to be eliminated; they are part of the character of interdisciplinary knowledge" (216). Such tensions emerged throughout the design timeline for *Vectors* projects, often occurring when a scholar was grappling with our database software and shifting her writing style away from the long-form essay. Such moments pushed scholars beyond their comfort zones and were likely experienced by them as a loss of expertise or as a deskilling. They became important opportunities to acknowledge the difficulties of collaborative work even as we worked toward easing that difficulty and completing the project. We also realized that it was important to let scholars know that such moments would occur from a project's outset, and to be alert to when they were likely to happen. Differences have also emerged between the core members of our lab as we have worked together over the years debating technology choices, aesthetic decisions, and timelines. The deliberative process and sometimes slow pace of the academy can be frustrating to those who typically work within corporate or freelance environments. Finding artists and technologists who were willing to tolerate the idiosyncratic nature of university life has been crucial to our work.

Tip Two: Value Experimentation and Process

If our design and technology collaborators needed to adapt in some ways to academic protocols, these team members also helped us to embrace experimentation and process. From books and articles published to grants and fellowships awarded, academic evaluation tends to focus on finished projects. The design practices in the *Vectors* Lab led us to appreciate process as an important element of the work we were undertaking and to embrace more experimental aesthetics. The first issue of *Vectors* went live in 2005, more than two years before the release of the iPhone and five years before the first iPad. Many of the pieces our team helped to create for *Vectors* were pushing the boundaries of existing digital scholarship, investigating the possibilities of digital screen language and multimodal expression for academic research. For instance, "Killer Entertainments" (2007) was published following a collaboration between feminist scholar Jennifer Terry and our designer Raegan Kelly. The piece re-contextualizes videos shot (often by soldiers) during the wars in Afghanistan and Iraq. These videos were shared on the web through sites like LiveLeak.com or YouTube (also launched in 2005) and were collected by Terry as documents of the conflicts and of emerging forms of visual communication. "Killer Entertainments" makes the videos a central part of the project, delivering the clips via a triptych design that asks the viewer to engage the video *as* video. Terry's analysis of the clips emerges as each video plays via a series of annotations and through a set of overarching themes and keywords that the viewer opens by clicking on small red circles that surround the triptych. Rather than use the videos as illustrations for a linear, textual argument, the videos are themselves incorporated into the piece, powerfully underscoring Terry's observation that "the combat camera has become a tool of evolving forms of entertainment for many who never find themselves in an actual armed conflict, and even for many who do." Terry's argument is meant to be experienced through multiple senses as the videos unfold and as the viewer navigates the piece. Video *matters* here along several registers, as do annotation and context. At a moment when the emerging web design industry was pushing for simple interfaces and "transparent" design, we embraced aesthetic experimentation as one route to a deeper understanding of emerging digital formats.

Projects like these were sometimes met with skepticism or confusion. They were not "easy" to engage. Each project assumed the form best suited to its argument and goals, and each required the reader or viewer to make sense of a new interface. While we may struggle with learning a new theoretical paradigm in print scholarship, the scholarly article and monograph are fairly recognizable as forms. *Vectors* pieces were challenging on many levels and especially so in their formal design. And, yet, this period of experimentation was crucial for our lab. Many of the projects seem less odd today after years of interacting with apps of many kinds and as video dominates the web. This design process also helped us to think through the shape that Scalar would eventually take. Rather than building technology for technology's sake, we built software to support scholars' work with the *Vectors* team. This software and the various scholarly projects produced for *Vectors* led to important insights about how Scalar should be designed. While many projects created with the platform are less aesthetically experimental than pieces created for *Vectors*, the platform also has an API that allows it to support a wide variety of design and aesthetic choices. The scholars with whom we have collaborated have also found that our workflow and methods shaped their interactions with their research materials. They have commented that the design process led them to new forms of writing, to new interpretations of their materials, and to expanded research methodologies. Many have continued to engage digital media practice, including, for example, Kimberly Christen. Following her time as a *Vectors* Fellow and her collaboration on the project "Digital Dynamics Across Cultures," Christen (2006) continued to work with our Information Design Director, Craig Dietrich, on software design projects. She has gone on to lead the team that created the culturally sensitive and ethically structured cultural heritage platform, Mukurtu. Platforms like Scalar and Mukurtu enable experiments in mediated and multimodal communication while also engendering rich forms of collaboration. They structure a valuable space for creativity and critique that exists in dialogue with more commercial forms, while also pushing beyond them. They help us to see that all technologies are situated within cultural contexts and that these contexts are vitally important.

Tip Three: Organize across Multiple Scales

Valuing experimentation is not without risks, and undertaking applied media research necessitates efforts at multiple scales. The academy is often slow to recognize emerging fields and non-traditional scholarship. Academics who carry out such work may find it difficult to explain how their research fits within the protocols in place within their departments, and scholars early in their careers are particularly vulnerable. I regularly review tenure files for scholars who have, in essence, undertaken twice as much research in order to be promoted, having published traditional monographs while also developing significant digital projects as well. My colleague Steve Anderson and I have each seen our paths toward tenure and promotion derailed and slowed down as a consequence of our commitments to applied media practice. Beyond our work on *Vectors* and Scalar, our lab is committed to joining larger conversations about the value of practice-based research within the university and to developing procedures for its evaluation. Anderson was the leading force behind the creation of USC's Media Arts + Practice doctoral program, devising curriculum that joined theory and practice and advocating for innovative dissertation formats. He and I worked together on a set of guidelines for the evaluation of multimedia scholarship, including an article published in *Profession* (Anderson and McPherson 2011), a journal of the Modern Language Association. Part of a special issue on assessing digital research, our essay developed the following ideas: (1) respect experimentation and emerging genres, (2) understand process, (3) appreciate collaborative and transdisciplinary approaches, (4) adapt current models of citation and peer review, (5) reward openness and appropriate contributions to the public commons, (6) value tools and infrastructure, and (7) remain flexible. I participated in the revision of my university's tenure handbook to account for collaborative and digital scholarship through service on USC's research committee. At the national level, scholarly societies such as the MLA, CAA, and AHA have released reports, offered workshops, and convened task forces to address evaluation procedures.

Such guidelines should reach far beyond considerations of tenure and promotion. As the number of tenure-track positions available at universities continues to decline, the ranks of contingent workers grow larger. This casualized labor often coincides with the staff that is integral to applied research or digital humanities projects. Our lab has largely been supported by external grants, a reality that continually threatens the long-term job security of our designers and programmers. While we have been fortunate to receive fairly regular funding to date, the pressure to write grant after grant can torque the research process in specific ways as researchers shape their work to fit the parameters of available funding. The existing reward structures within the university tend to favor tenured academics and often make it difficult to include non-academics as principal investigators on grants. While all of the work within our lab has been collaborative and neither Scalar nor *Vectors* would exist without the committed labor and creative insights of Erik Loyer, Craig Dietrich, Reagan Kelly, Curtis Fletcher, Steve Anderson, and many others, I am often seen as the public "face" of our work. For applied research in media studies to develop in ethical and just ways, such inequities have to be openly discussed and fairly resolved. There is a very real risk that our zeal for applied work will inadvertently dovetail with larger forces that increasingly corporatize the university. In building a space within the university for digital scholarship that encompasses feminist and progressive themes, we should also be mindful that we are not undermining the possibility for better working conditions for many of those who join us.

Gary Hall (2013), David Golumbia (2014), and others have argued that the approaches favored within some applied digital research may be incompatible with the values of the interpretative humanities. While I do not fully agree, I do believe it is important that we take such claims seriously and consider the harm that might result from a turn toward digital scholarship. For instance, in addition to grappling with labor issues, our lab continues to struggle with best practices for ADA compliance in our software design. We also worry that our push toward open access publishing might inadvertently put the burden on scholars to take on more and more of the labor for producing their publications, devaluing specific expertise that exists within university presses, archives, and libraries. While it is tempting to work on individual projects and to develop our platform apart from these broader concerns, our work exists in a larger ecosystem that demands our attention to many scales.

Tip Four: Cultivate External Relationships

Our early work on *Vectors* was more internally oriented, even as we brought in fellows and ran summer institutes.[4] We were focused on the critical and theoretical questions that motivated the scholars with whom we worked, humanities scholars investigating questions of memory, race, gender, embodiment, sexuality, perception, temporality, ideology, and power. We were interested in seeing how you might immerse yourself in an article in multiple affective and sensory registers. We wanted to see if you could play an argument like you might play a game. We asked what might happen if scholarship explored the emerging vernaculars of the digital, drawing from both artistic and popular expression. We were not initially focused on how our work would intersect with libraries, presses, and other institutions, but we were quickly drawn into these conversations. Soon after the journal launched, I was invited to speak at the Coalition for Networked Information, and this began a series of useful and thought-provoking conversations with librarians and technologists. Librarians pressed our team to think more robustly about questions of preservation and sustainability. They asked why we were not developing templates and worried that our projects would not be accessible far into the future. If it was important for our lab's work to begin in a place of largely unfettered experimentation, it also became clear that our efforts needed to be in dialogue with other related projects. We needed to find a balance between experimentation and sustainability. Our team members spent an increasing amount of time in dialogue not only with our fellows and other designers but also with librarians, archivists, foundation officers, technology developers, journal editors, deans, provosts, and publishers.

These external conversations became a crucial context for the development of Scalar. The work on *Vectors* had been incredibly fulfilling, but the fellowship process was expensive and time consuming and would be impossible to fund forever. We were seeking to build upon that research and find ways to scale and extend it, making aspects of our process more widely available, particularly to scholars at under-resourced institutions where such fellowships would not be possible or where technological infrastructure was lacking. With support from the Mellon Foundation and the National Endowment for the Humanities and in close collaboration with many colleagues (including especially Wendy Chun, Brian Goldfarb, Nicholas Mirzoeff, and Joan Saab), we began a larger organization, the Alliance for Networking Visual Culture, that became the home for Scalar. Drawing upon lessons learned through *Vectors*, we aimed to support new ways of working with digitized archival materials within the humanities and to continue to push for new modes of digital scholarly publishing. In the context of specific research questions in visual studies and media studies, the Alliance focuses on integrating the primary source materials available in online databases more directly into born-digital scholarship. We located our technological infrastructure, Scalar, within a larger institutional ecosystem that included partnerships with archives and museums (including the Shoah Foundation, Critical Commons, the Hemispheric Institute's Digital Video Library, the Internet Archive, and the Getty) and with several university presses (including MIT, California, Duke, Stanford, NYU, Cambridge and the Open Humanities Press) to create a testing ground for the investigation of new publishing possibilities. We also have worked closely with scholarly societies like the College Art Association, with several humanities centers and university libraries, and with other digital publishing projects. Recent partnerships include a burgeoning connection to the Los Angeles Unified School District.

These partnerships are important, but they are often challenging. If our early design process for *Vectors* demanded that we build a shared vocabulary across different fields within a small team, these collaborations considerably intensify those needs. The issues faced by different types of partners do not always align. For instance, librarians' focus on preservation can be at odds with a scholar's desire for innovation. University presses may be less invested in open access scholarship than many libraries. Archives' priorities may be centered more on digitization than on publication. Scalar was conceived as a platform that could help university presses ease into digital publishing. While we have published a large number of Scalar pieces with a variety of presses, our impact on presses remains less robust than we had originally hoped, and we have found it challenging to get presses to collaborate with each other. We sometimes joke that the code is easy, but the people are not. The benefits of applied media studies will not be achieved through technological solutions alone. Media and film scholars should be working with designers, technologists, presses, policy makers, lawyers, librarians, archivists, and activists both to create our research and to shape the networks in which it circulates.

Tip Five: Expect the Unexpected

As we have experimented with both scholarly projects and scholarly ecosystems, we have also come to expect the unexpected. Unanticipated moments can lead to positive outcomes, but, at the time of their occurrence, they often do not feel welcome. Take the example cited at this essay's opening. When we were contacted by the young women who felt that a public Scalar project violated their privacy, we needed to make a judgment call about whether or not to shut down the piece. Upon close inspection of the project, it did not seem that the author had disclosed her presence when interacting with private pro-anorexia online communities. The piece thus seemed to be in violation of best practices for online ethnographic research and also to be in conflict with Scalar's Terms of Use. Before shutting the project down, we reached out to the author, who replied that she had created it for a graduate course and was fine with it being taken offline. Thus, this particular episode was resolved fairly quickly, but the incident led to a period of reflection for our team. What does it mean to make available an open, online platform that can potentially be used in ways that might not line up with the ethical commitments of our team? How does one weigh the rights of an individual author against the integrity of a platform developed in a scholarly setting? Such questions are not abstractions. As commercial platforms like Twitter, YouTube, and Facebook grapple with the circulation of racist, sexist, and violent content through their networks, ethical questions about technological development abound. While Scalar obviously has a much smaller footprint than these social media giants, our setting within an academic environment makes such questions very real and pressing. Before launching Scalar, we crafted our Terms of Service with the pro bono advice of American University's intellectual property legal clinic, striving to use clear, accessible language, a process spearheaded by Steve Anderson. These terms give us broad latitude to monitor content published on our servers and include the following:

> Users agree not to use the Websites or Services to [p]ost or transmit Content that is harmful, offensive, obscene, abusive, invasive of privacy, defamatory, hateful or otherwise discriminatory, false and misleading, incites an illegal act, or is otherwise in breach of your obligations to any person or contrary to any applicable laws and regulations.
>
> *http://scalar.usc.edu/terms-of-service/*

These Terms of Service allow us to restrict content in Scalar projects, particularly those on our servers, but such broad guidelines become real when specific conflicts arise. This particular incident propelled our team to discuss various "what if" scenarios in much more detail than we had previously done, and to clarify for ourselves our own understanding of "hateful or otherwise discriminatory" content.

Other unanticipated issues have included various technological breakdowns and complications. A very popular and well-received *Vectors* project is Emily Thompson's *The Roaring 'Twenties*, an interactive exploration of the historical soundscapes of New York City. It richly draws from the Municipal Archives of the City of New York, cataloging over six hundred unique complaints about noise around 1930 while reproducing over 350 pages of these materials. It also includes fifty-four excerpts of Fox Movietone newsreels as well as hundreds of other photographs and print materials. A visitor to the piece can navigate along three pathways that organize the site's content via the rubrics of space, time, and sound, each offering a unique vantage point on the database that drives the project. *The Roaring 'Twenties* was originally built with the Adobe program Flash, layering a historical map of New York on top of a Google map for the spatial section and embedding various media files. About a year after the project launched, we received a panicked message from Emily that the piece was not working. Following various industry machinations meant to throttle Flash (for instance, Apple chose not to support Flash in iOS), Google Maps changed its support for the platform as well, rendering *The Roaring 'Twenties* inaccessible. The piece had to be reconfigured to work with MapQuest, not a trivial task. Many other *Vectors* projects were built in Flash and may prove inaccessible in years to come, illustrating the difficulties inherent in preserving digital scholarship. Another morning found all of Scalar inoperative, leading to a series of phone calls with the USC division that supports and backs up our servers. Some sleuthing determined that our university uses an outside company to scan all of its servers for obscene or illegal material. The content of a Scalar project had triggered an automatic shutdown of the entire site because an initial warning went to an incorrect email address. Such software helps us monitor Scalar for offensive or racist content, but it also can trigger other problems. My doctoral degree in feminist film studies did not really prepare me to manage such infrastructural meltdowns. Each incident helps us to refine and rethink our policies and procedures, but they also can lead to tense moments and wasted time.

There have also been more positive "unexpected outcomes." While we designed Scalar as a scholarly authoring tool to help address certain issues in academic publishing, we have been delighted to see the platform broadly taken up for classroom use. Several university libraries and technology centers now support Scalar for pedagogical purposes, and we continue to hear good feedback about how Scalar is being used across the curriculum for multimedia essays, collaborative authorship, archival engagements, and theses and dissertations. For instance, a freshman seminar at the University of Pennsylvania in the fall of 2015 used Scalar as a collaborative workspace for students to collect, analyze, and publish their work with nineteenth-century archival materials across a whole semester.[5] Their project pushed Scalar in some interesting new directions that led us to expand the platform's feature set. Now that we have seen a wide range of educational projects undertaken with the platform at the undergraduate and graduate level, we are beginning to explore the use of Scalar in middle and high school environments through a partnership with the Los Angeles Unified School District.

Our lab's fifteen years of work in applied media studies has certainly not been without tensions both small and large. Our collaborative and iterative methods of working are often messy and entangled. Such processes bring scholars, technologists, librarians, publishers, artists, archivists, activists, software, hardware, networks, media objects, and larger institutions into shifting, sometimes unstable, assemblages. These formations can surface gaps in various vocabularies and in modes of working that can be challenging to negotiate, but they can also produce new configurations of knowledge and forge connections across and beyond the university. As a media scholar, such work continually reminds me that our technologies are never neutral. Of course, such a sentence is almost a truism of media and cultural studies. Applied media studies can move such a statement beyond abstraction toward a deepened awareness that our technological systems are not entirely beyond our reach. Designing technological systems helps us better understand how these systems work, as well as how they work on us, or, as Miriam Posner (2015) has observed, we can better see "how provisional, relative, and profoundly ideological is the world being constructed all around us" through the digital. But that is not all. Having collaborated within our lab for many years to imagine and build online projects and open publishing platforms, I also know that we can—at least now and again—bend technology more closely in line with our desires for a better world.

Notes

1 For more information on the various collaborations that have supported *Vectors* and Scalar, please see the websites for the two projects, http://www.vectorsjournal.org and http://scalar.usc.edu. Across the years, many individuals and institutions have partnered with us or otherwise contributed to our work.
2 A notable exception was the journal, *Kairos*, an early pioneer in multimedia scholarly publishing, as well as experiments such as *Horizon Zero* from Banff and the *Electronic Book Review*. Find them at http://kairos.technorhetoric.net/, http://www.horizonzero.ca/, and http://www.electronicbookreview.com/.
3 Information about The Labyrinth Project is available at https://dornsife.usc.edu/labyrinth/laby.html.
4 We did undertake an extensive planning process as we developed the initial contours of the journal and its fellowship process. This included hosting a summit with several leading scholars, including the directors of noted humanities centers. But, in retrospect, such a session might have usefully included librarians, technologists, and designers.
5 The project was spearheaded by Ian Petrie, the Senior Associate Director of the Center for Teaching and Learning at Penn. It is available at http://scalar.usc.edu/works/the-voyages-of-the-clarence.

References

Anderson, Steve, and Tara McPherson. (2011) "Engaging Digital Scholarship: Thoughts on Evaluating Multimedia Scholarship," *Profession*, 136–151.

Balsamo, Anne. (2010) "Design," *IJLM* 1(4): 1–10.

Christen, Kimberly, and Chris Cooney with Alessandro Ceglia (designer). (2006) "Digital Dynamics Across Cultures," *Vectors* 2(1). www.vectorsjournal.org/projects/index.php?project=67

Dourish, Paul. (2001) *Where the Action Is: The Foundations of Embodied Interaction*. Cambridge, MA: MIT Press.

Golumbia, David. (2014) "Death of a Discipline," *Differences* 25(1): 156–76.

Hall, Gary. (2013) "Towards a Post-Digital Humanities: Cultural Analytics and the Computational Turn to Data-Driven Scholarship," Special Issue on New Media and American Literature, Tara McPherson, Patrick Jagoda, and Wendy H. K. Chun (eds.), *American Literature* 85(4): 781–809.

Klein, Julie Thompson. (1996) *Crossing Boundaries: Knowledge, Disciplinarities, and Inter-disciplinarities*. Charlottesville, VA: University of Virginia Press.

Posner, Miriam. (2015) "What's Next: The Radical, Unrealized Potential of Digital Humanities." http://miriamposner.com/blog/whats-next-the-radical-unrealized-potential-of-digital-humanities/

Terry, Jennifer with Raegan Kelly (designer). (2007) "Killer Entertainments." *Vectors* 3(1). www.vectorsjournal.org/projects/index.php?project=86

5

THE MEDICAL FUTURES LAB

An Applied Media Studies Experiment in Digital Medical Humanities

Kirsten Ostherr

While debates about how to handle medicine's digital transition have intensified in recent years, the use of computers in medicine has long raised questions of how to define appropriate boundaries between human and technological roles in healthcare. Arguments about the therapeutic power of human touch (Verghese 2015) collide with shocking statistics on the thousands of adverse events—including deaths—caused annually by human error in medicine (Kohn, Corrigan, and Donaldson 2000). Starting with the Health Information Technology for Economic and Clinical Health (HITECH) provisions of the American Recovery and Reinvestment Act of 2009, healthcare reform incentivized clinicians to adopt electronic health record (EHR) systems (CMS 2015). Doctors, nurses, and patients have been complaining about the poor usability of their new systems ever since. Demand for evidence-based, personalized care is driving increased adoption of artificial intelligence systems in healthcare, exacerbating fears of the dehumanization of medicine (Crawford and Whittaker 2016). Mobile technologies are moving care outside of formal clinical settings and placing greater opportunities—and responsibilities—in patients' hands (Chan et al. 2012; Greene 2016). Everywhere we look for healthcare, we find electronic screens, digital interfaces, and bad design. In this context, the need for human-centered technologies that respond to patients as people, not merely as data points, only grows more pressing. The domain of medicine is clearly full of opportunities for humanists to contribute to the design and social intermediation of digital interfaces that could improve healthcare experiences for patients and for doctors. What is less clear is how to go about doing that. In what follows, I describe our efforts at the Medical Futures Lab to develop applied media studies methods for bringing health humanist insights to health information and communication technology (ICT) design.

This chapter describes an applied media studies project that emerged in response to a set of conceptual problems surrounding the role of technology in doctor-patient communication. By narrating the interests and actors involved in the development of the Medical Futures Lab, I show how applied media studies can lead from theoretical problems in human-computer interaction to practice-based answers in technology design. In the first section I describe an improvised experiment in collaborative applied media studies that provided early insights about finding good partners, and formed the foundation for larger projects. The second section narrates the ideation process that led to the formation of the interdisciplinary medical humanities media lab called the Medical Futures Lab. In the third section I describe the "tooling up" process of acquiring translational skills for collaboration across different fields, while the fourth section details the specific projects that we have undertaken at the Medical Futures Lab. Section 5 discusses our efforts to create a reflective, discussion-centered medical humanities MOOC called "Medicine in the Digital Age." The sixth section places this work in historical context by taking the long view on debates about technology in education, framing those discussions in relation to concerns about computers in medicine. The chapter closes with a concise list of the building blocks and biggest challenges to doing applied media studies work at the intersection of humanities and medicine, followed by two sets of advice: one for experimenters, and one for the administrators that support them.

Phase One: Project TMC

I begin with a story about the people who came together to do this work for two reasons. First, I want to emphasize the importance of the individuals involved in creating from scratch a multi-institution, multidisciplinary, collaborative lab to solve problems in digital health. Humanist media scholars already know that human contexts of use shape the meanings we make from technologies of representation. It is also true that human contexts for doing creative, critically reflective hands-on work shape the design that comes out of those experiments. That is, finding good partners is vitally important, and while there are some useful operating principles one can follow, there is no generic formula for getting it right. I narrate the story of how I found my collaborators to illustrate this process. Second, I want to emphasize the role of the main characters in shaping the course of the story as it unfolds. How best to approach an undertaking like creating the Medical Futures Lab was not transparently obvious. Imagining a humanities-based lab that operated at the intersection of tech startups, research universities, medical schools, and media design firms, I had big aspirations but modest financial and human resources. Telling the story of the people who created the Medical Futures Lab illuminates how their unique affinities, interests, and resources have played a decisive role in what we do and how we do it.

The story begins with a party. In the summer of 2010, I attended a banquet to celebrate a major milestone for the McGovern Center for Humanities and Ethics at the University of Texas Medical School in Houston. There I met another guest, David Thompson, who was a founder and principal at ttweak, a multidisciplinary communications and design firm. Thompson specialized in digital media campaigns for healthcare initiatives and had recently launched a major marketing campaign for a children's hospital in Houston, so he was thoroughly immersed in conversations within the Texas Medical Center (TMC) about how representation through digital storytelling shapes our understanding of what medicine is, and what it can do. As we talked, I heard a vivid description of how ttweak used different forms of visual imagery and storytelling to shape doctors' and patients' experiences in healthcare settings.

I want to emphasize here that as a media studies scholar, my stance toward advertising and digital marketing practices was a critical one. I regularly taught my "Medicine and Media" students how direct-to-consumer pharmaceutical advertisements invented new medical conditions to sell more drugs (Dumit 2012). I critiqued the commercialization of medicine as a field of practice. I was not looking to partner with a marketing firm. Yet, in my conversation with Dave Thompson, the opportunity was suddenly obvious: what if students who learned about technologies of representation in medicine could apply their knowledge by working on a creative project that involved some form of media production for a healthcare context? Inspired by the principles of hands-on, active learning-by-doing that characterize many applied media studies projects (though we were not using that term at the time), Dave and I began to brainstorm ways to collaborate on a project that would bring our different approaches to medicine and media together.

The role of serendipity in innovation has been well described (Hagel et al. 2010), and its role in my initial encounter with my soon-to-be-collaborator was plainly evident. Our presence at that party was not completely random—we were both connected to the host. But no one introduced us or ensured that our potential as collaborators would surface in conversation. Fortunately, three key signals did appear early on to set our experiment in motion: we had complementary but different skills, we enjoyed each other's company, and we were both willing to take risks on creative projects. Equally important discoveries came later: we were both able to remain flexible and adaptable in the face of changing circumstances, and we were both capable of failing (more or less gracefully) and persevering with humility and humor. I must add that I have learned in the years of building the Medical Futures Lab that preconceived notions about who brings what to the table are often misleading. Assumptions about the vested interests of actors in adjacent fields, including marketers and pharmaceutical representatives, are not always accurate and do not usually tell the whole story about the people in those roles. In translational contexts, listening with open ears is an essential practice, and the true value of critique lies in opening a path to productive, collaborative intervention.

Johnson (2010) and others have discussed the importance of collaboration through difference in fostering innovation, and Davidson (2011) has explained the value of that approach for reimagining higher education in the twenty-first century. Inspired by these ideas, we began a collaboration that started as a side project embedded within my "Medicine and Media" class. We called it "Project TMC," and recruited a small group of students to participate in multimodal, multimedia ICT creation in partnership with a few clinicians at the Texas Medical Center (TMC). As we explained to potential recruits, "By bringing students trained in literary and visual practices of interpretation and representation together with clinicians and researchers, we aim to present a new perspective on the vital interconnections between humanistic methods of engagement and scientific practices of healing" (Ostherr and Thompson 2010). Our hypothesis was that the students' interest in medicine, coupled with the knowledge they gained through class readings and discussions, would allow them to bring creative insights to their framing and visualization of the "behind-the-scenes" world they observed in the medical center. Our goal was to deepen the students' critical humanist approach to analyzing and interpreting medical visualizations by engaging them in hands-on critical and creative work (Rheingold 2012).

We selected three students to participate in the pilot project. We figured out how to give a crash course in digital storytelling, how to translate concepts from class readings—such as "the myth of photographic realism" or "the anatomical gaze"—into the students' creative design practices, and how to explain to our clinician partners what these students were doing and why. The students worked on two projects: helping to translate information about a complex medical condition to pediatric caregivers using a blend of text, clinical images, and animation, and participating in the professional production of a documentary film about the Surgeon-in-Chief at a local children's hospital.

We gained several key insights from this experiment. First, having a tangible, clearly defined goal (or end product) is critical for facilitating collaboration in diverse teams, but the educational impact of such projects is more deeply connected to the process than to any of the resulting artifacts themselves (McPherson 2009). For the students, the experience of working creatively to help a clinician solve a real problem and benefit patient care was enormously valuable. In contrast, our clinician partners cared most about the outcome of the process. As educators themselves, they were pleased to contribute to the pre-med students' training, but as busy clinicians, they wanted tangible results. These realizations about the importance of reciprocity and setting expectations played a critical role in the transformation of Project TMC into the Medical Futures Lab.

Phase Two: Imagining the Medical Futures Lab

The pedagogical power of the hands-on, applied media studies approach was evident from the start of Project TMC. For the next phase, I wanted to scale up and develop projects with a broader network of students and collaborators, so I began to search for potential long-term partners. At the time, I sometimes described my vision as "an MIT Media Lab for medicine." What I wanted were thinkers and makers from the diverse fields that constituted the changing digital health landscape: medical doctors, computer scientists, engineers, visual artists, start-up founders, and other experimentally engaged humanists.

I began by drafting a brief vision statement that I could share with potential collaborators. I called it "The Medical Futures Initiative," and gave it the tagline, "Training the medical media innovators of the future through creative, hands-on critical thinking and design." In this statement I tried to capture the aspirational, collaborative, and interventionist approach that we had begun to test out in Project TMC. The longer description stated,

> Medical futures is an experiment-based educational laboratory for retooling how doctors are trained in a digitally interconnected age. We focus on how new ways of creating and sharing knowledge and data through digital media change the ways doctors communicate, conduct research, and practice, just as they change the ways patients engage with medical professionals. Our goal is to find new ways of imagining what we can do with the new media tools at our disposal in the service of better medicine and health.
>
> *(2011)*

This document was intended to serve both as a focal point for my own efforts and, in some sense, as a recruitment tool for collaborators, so I went on to describe the attributes I was seeking to find and cultivate through this project:

Core Values:
Applied and Engaged Humanism
Anti-Disciplinarity/Anti-Silo
Hard Fun and Serious Play
Immersive and Experiential Learning
Collaboration through Difference

Core Objectives:
Foster a creative environment for imagining what the future of medicine *should* be.
Create media and other tools for improving healthcare and medical education.
Teach pre-med students to think analytically and historically about medicine, science, and technology.
Develop new tools for posing novel questions that connect teaching, research and practice.

The last section of the proposal sketched out the components and timeline for bringing this vision to fruition, including development of new courses to be co-taught with TMC clinicians at Rice University, a symposium, and new courses to be co-taught at local medical schools on topics such as "Computational Thinking for Doctors." Finally, I listed several long-term future goals: recruiting and training graduate students and postdoctoral fellows; collaborative creation space housing humanists, designers, physicians, scientists, and others under one roof; and medical clerkships in key fields of digital health.

While I was already meeting with potential collaborators in the visual arts, computer science, and many other academic fields at Rice University, I knew that because Rice does not have its own medical school, I would need to reach beyond my home institution to identify clinician partners. In proper twenty-first-century fashion, I looked online, and at the time, the first hit for a "medicine social media" Google search was a blog called "33charts." The author of the blog, launched in 2009, was identified as Doctor Bryan Vartabedian, and he was fortuitously located just down the street in the Texas Medical Center. I found his Baylor College of Medicine email address, reached out to him, and told him briefly what I was hoping to do. It was a cold call, but Bryan's insightful blog posts about medicine, social media, technological change, and the value of public thinking suggested that he might find the idea of the Medical Futures Lab intriguing. My message piqued Dr. Vartabedian's interest, so we met and discussed. This marked the beginning of a critically important collaboration that would define the shape of the Medical Futures Lab.

As we brainstormed ideas for the Medical Futures Lab, Bryan and I focused in on the need for critical public dialogue and teaching to address what we saw as a major gap in medical education. In our experiences as a college professor and a practicing clinician, we saw that pre-med students lived in a digitally intermediated world, where social, educational, and extracurricular activities all took place through a complex blend of participation in online and offline spaces that, in our students' view, were seamlessly interconnected. For these students, the distinction between "online" and "offline" was irrelevant, as their social lives moved fluidly across these domains (boyd 2014). When presented with a group assignment, our students would immediately establish a collaborative digital workspace by creating a Facebook group, a shared Google document, a GroupMe chat, and other forms of connection as needed. They would interact through all of these platforms both synchronously and asynchronously, even when they were physically in the same location, sitting face-to-face around a table. Our pre-med students' use of information and communication technologies (ICTs) was unremarkable to the students themselves; the tools were ubiquitous parts of their lives that were familiar from many years of use. Yet, once these students entered medical school, those tools were sealed off from their training by policies that treated connected learning like a HIPAA violation, and Facebook like a lawsuit waiting to happen.

In light of this disconnect, we focused on two major areas of concern that would shape the work of the Medical Futures Lab. First, we noted that despite our students' ease in adopting new technologies, they lacked a critical framework for assessing the broader cultural, political, economic, or ethical implications of their activities on the mobile, social web. Second, we saw that our students gained real social, creative, and collaborative benefits from their use of these tools (Jenkins et al. 2006), only to lose those assets when they entered their training as medical professionals (Baruch 2017).

The absence of a critical framework for reflecting on the implications of digital ICTs meant that our future health professionals were disadvantaged in their ability to assess appropriate forms of social media engagement in their capacity as doctors. Failure to cultivate critical reflection on the tools of the social web in health professions training meant that educators were alienating their students, diminishing an important source of social support, reducing their ability to perform at their highest level, and shutting down the potential for innovation through learner-centered pedagogy. Worse still, this meant lost opportunities to interact with patients using tools that both doctors and patients prefer in their non-clinical lives, potentially curtailing new forms of engagement that could lead to improved health outcomes, reduced cost, and better care (Kvedar, Coye and Everett 2014). Finally, this gap in their education meant that our future doctors would move into the increasingly techno-mediated world of medicine without the critical skills to assess when asynchronous digital exchange was an appropriate form of information sharing and communication, and when synchronous distant or in-person interaction was necessary.

To address these gaps, we created the Medical Futures Lab as a multidisciplinary collaboration space that would engage in research and teaching on emerging digital health areas of concern, including the digitally intermediated doctor-patient relationship, social media in clinical settings, the quantified self in the exam room, big data ethics, and peer-to-peer healthcare. Our goal was to cultivate a practice of creative, critical thought that we called "digital medical humanities." A founding principle was that, in the age of user-generated content, there are no more passive consumers of information. Therefore, twenty-first-century doctors (and patients) needed to cultivate new competencies, including content creation, curation, and sharing to facilitate transmedia storytelling that could capture patients' lived experiences of health, illness, ability and disability beyond the limited context of the clinical encounter.

Phase Three: Tooling Up

We knew the problems we wanted to address, and we knew the competencies we wanted to cultivate. What would it take to do this kind of work? While my long-standing involvement in film and media studies meant that I had participated in several small-scale filmmaking projects throughout my life, that kind of "hands-on" work played a minor role in fostering my ability to do this kind of work. More important, I believe, was my exposure to diverse creative settings such as digital health startups, hackathons, and human-centered design sprints, and my training in public health. Part of my motivation to create something like the Medical Futures Lab came from my experience translating my research on historical health and medical films to audiences of health professionals. In my lectures, I worked to show groups of medical doctors and public health experts how and why the aesthetics and contextual framing of visual representations mattered to them and their patients. These audiences were receptive, and they expressed their sense that this was an important gap in their training that was becoming more pressing as medicine became more digitally intermediated. I realized that the field of humanistic media studies rarely interacted with medical and public health practitioners whose work entailed image creation and interpretation. To develop the methods and cross-domain fluency to expand my own work at this intersection, I returned to school and completed a Master of Public Health (MPH) degree with the support of a Mellon Foundation New Directions fellowship.

My MPH training played a critically important role in shaping the work we would do at the Medical Futures Lab in two specific ways. First, I was exposed to valuable new approaches to innovation, health communication, and intervention design in several of my courses. Learning new methodologies expanded my capacity to think creatively about new collaborative projects at the intersection of humanities and medicine. Second, I became fluent in the language of health promotion and behavioral science research. By diving deep into the literature, attending classes, writing papers, and taking exams, I learned how health professionals frame and explore information and communication problems, I identified the critical gaps in research on digital health ICTs, and I gained direct experience in translating humanist media studies questions into applied forms of research. When I finished my degree, I could speak health communication research in a way that opened up new kinds of collaboration with clinicians who shared my interests but described them in what had previously been a foreign language to me. The immersive experience of a multi-year degree program may not be strictly necessary for every kind of translational applied media studies project, but in cases where the focus is on bridging two distinct fields of expertise, the credentials are invaluable.

In the case of the Medical Futures Lab, the need for tooling up one's skills to become adept at translational creative practice is not limited to expanding on humanist methods. Our core clinician partners have also dedicated extensive effort to becoming bilingual. For instance, in addition to his practice as a pediatric gastroenterologist, Bryan Vartabedian spent many years honing his public communication skills through his blog, where he translated his clinical expertise into reflection on the broader landscape of digital health transformation. Peter Killoran brought his background in geography and software development to his practice as an anesthesiologist and biomedical informaticist.

In merging those fields, he developed unique translational expertise in the practices of transforming narrative into data, and data into visual display. In the development of applied media projects, translational abilities are often more important than technical skills.

Phase Four: Creative, Hands-on Critical Thinking and Design

With the support of an internal Rice University grant, the Medical Futures Lab launched our first round of activities by designing and teaching collaborative courses in partnership with medical doctors. Each course involved hands-on critical-creative assignments, such as a project I developed with Peter Killoran, MD, to redesign electronic health record systems using the principles of patient engagement and narrative medicine. With Bryan Vartabedian, MD, I developed and taught "Medicine in the Age of Networked Intelligence," a new class where we incorporated extensive hands-on media assignments ranging from blogging to video creation, all posted on a public course Tumblr site. These courses aimed to fill the gaps in medical and pre-medical training that we had identified in the lead-up to founding the Medical Futures Lab. Our approach included exposing our students to the current debates in digital health, describing the state of technology development, helping them cultivate the skills to ask critical questions about emerging health technologies, and fostering digital literacies through hands-on practice and critical reflection. Assignments were designed to give the students direct experience with online content creation, curation, and sharing to facilitate transmedia storytelling in health contexts (Figure 5.1).

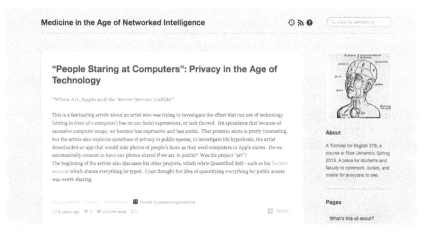

FIGURE 5.1   Course Tumblr page for "Medicine in the Age of Networked Intelligence," April 24, 2013.

Source: Photo by Kirsten Ostherr.

To extend our collaborative network, the Medical Futures Lab hosted a symposium, "Millennial Medicine: Knowledge Design for an Age of Digital Disruption" that brought together a mix of digital health innovators such as Eric Topol and Marc Triola, visual artist Alexa Miller, open source code pioneers Fred Trotter and Rich Baraniuk, applied media humanist Anne Balsamo, and others. We pursued our aim of fostering public dialogue around the challenges and opportunities in digital health through this event, other public presentations, a Medical Futures Lab blog, and participation in the local health startup community, particularly the local chapter of the national organization called "Health 2.0."

The next major development of the Medical Futures Lab fused the objectives of generating public dialogue and cultivating critical and creative digital skills. We created the Medical Media Arts Lab, a collaborative, participatory design class that brings Rice undergraduate students together with physician mentors from the Texas Medical Center to solve real problems that clinicians face in their work. Our inspiration for developing the Medical Media Arts Lab came from numerous sources, including Project TMC, humanist scholar-technologists (Jenkins et al. 2006; Balsamo 2011; Davidson 2011), media lab leaders (Moss 2011; Maeda 2013), and design thinkers (IDEO 2011), as well as creative/critical practitioners from more proximal environments, such as architecture studio critiques and engineering design kitchens at Rice. The Medical Media Arts Lab follows a process of human centered design shaped by readings on media theory, information and communication technology (ICT) research, digital health challenges and opportunities, and health policy. Teams of students work with a clinician "problem owner" for fifteen weeks to design a solution to a complex, but clearly defined health communication or visualization problem. The process includes deep end-user research, brainstorming with problem owners, participant observation, storyboarding, prototyping, and other techniques of experimental, collaborative design (Figures 5.2 and 5.3).

The design problems have dedicated mentors and specified parameters, and faculty provide guidance, feedback, technical skills, access to relevant environments for observation, and contact with key people to interview. Students have worked on timely, pressing issues such as patient enrollment in clinical trials for personalized cancer drug therapies, goal-setting communication between doctors and patients with diabetes, and improving caregiver engagement on cardiovascular intensive care unit (ICU) rounds. Projects have included production of short videos, graphic design, infographics, virtual three-dimensional models, game design, and app development. I have described in detail elsewhere the process for running the design labs (Ostherr 2018).

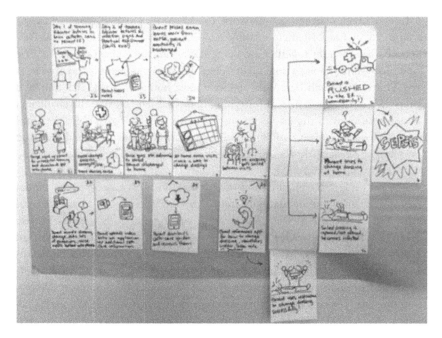

FIGURE 5.2 Team sepsis storyboard for "Medical Media Arts Lab," Spring 2017.
Source: Image Courtesy of Abby Halm, Linda Lin, Christine Luk, Mia Schmolze, and Chad Zhao. Photo by Kirsten Ostherr.

An important feature to emphasize here is that the students engage in extensive hands-on applied media practice, but they do not receive extensive direct training to do so. The student teams are purposefully assembled to include a mixture of backgrounds and talents, but even when a team does not have existing skills in digital content creation, they receive just enough training to feel confident that they can figure it out on their own. Confirming our sense that our current students (and future doctors) find that many emerging media configurations have intuitive usability, our students happily experiment with new tools to get their work done. Also, as expected, most students lack the critical framework for assessment and reflection on these tools, and that is what the class provides. Rather than spend extensive class time teaching techniques for how to use specific tools, we spend time discussing how to figure out what is the best tool for a given information or communication problem, and once selected, what questions to ask to further assess the broader social and cultural dimensions of the technology's affordances. We also spend time developing critical skills for online information filtering, so that the teams can independently locate and evaluate reliable sources.

FIGURE 5.3 Team e-Patient: Alex Lam, Obi Nwabueze, Mollie Ahn. "Medical Media Arts Lab," Spring 2014.

Source: Photo by Kirsten Ostherr.

Phase Five: Connected, Online, Hands-On—the Digital versus the Analogue

Our work with millennial students, educators, clinicians, patients, and technologists has taught us over and over again that the categories of "online" and "offline" have diminished relevance in the context of twenty-first-century higher education, clinical practice, and non-clinical patient experience. Yet enormous gaps remain between "online" and "offline" experience in clinical training and formal patient care settings. We continue to find a real need for critical thinking, dialogue, and experimentation that explores what the boundary between the human and the technological is and should be. When does a patient need face-to-face, physically co-present interaction with a physician, and when would a remote, asynchronous video- or text-based exchange suffice, especially if it saved the patient a time consuming and expensive trip to the clinic? What are the irreducibly human elements of caregiving that artificially intelligent machines cannot replace? How might we design an EHR interface that is as pleasurable and easy to use as the most popular social media platforms? This sampling of questions points to the broad range of issues needing ongoing teaching and research at the intersection of medicine, technology, and hands-on creative practice.

The field of medical education has been slow to update its pedagogical strategies for millennial learners, despite calls for change (Prober and Heath 2012; Topol 2013). Consistent with our perspective at the Medical Futures Lab, some experts have noted that critical thinking skills related to digital literacies will become essential skills for future medical professionals (Chretien et al. 2015). To meet this need, we developed and taught a massive, open, online course (MOOC) for medical audiences using humanistic instructional design methods to encourage dialogue and engagement with concepts (rather than solely conveying factual information). Why make a MOOC? Returning to our earliest conversations about forming the Medical Futures Lab, we saw that our efforts might be filling a gap locally, but very few medical schools had the interested faculty or the willing administrators to get novel content and methods around digital health into the curriculum. We saw the free, open-access, asynchronous, and globally available MOOC model as an opportunity to help build the field and make our creative-critical thinking available to a much broader audience.

Based on a previous face-to-face Medical Futures Lab course we had taught at Rice ("Medicine in the Age of Networked Intelligence"), Bryan Vartabedian and I developed a MOOC called "Medicine in the Digital Age" which ran on the edX platform. The course was developed over eight months during 2014–2015, and it went live in May of 2015 for a six-week course run. Over 15,000 students have taken this course since it launched, and we have engaged in dialogue with participants from around the world, including physicians and patients in India, pharmacists in Iran, caregivers of aging parents in Mexico, and many more. Drawing on the vast datasets of participant contributions to

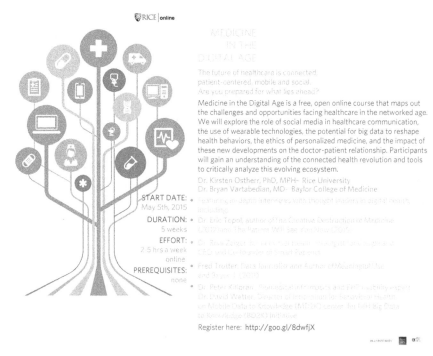

FIGURE 5.4 "Medicine in the Digital Age," MOOC poster. Spring 2015.
Source: Used with permission by Rice Online.

the online discussion forums in the course, we are working on research with colleagues in computer science that explores whether it is possible to identify and quantify "humanistic knowledge formation" in this large-scale, online medical humanities course (Figure 5.4).

As the Medical Futures Lab has grown, interest in doing "applied," engaged, or experiential research and teaching has been embraced by many fields within and beyond higher education. The value of learning through hands-on practice is neither new nor unique to applied media studies or digital medical humanities. For the past decade, research from neuroscience, psychology, and other fields has emphasized the importance of active, participatory learning across all stages of education, fueling critiques of traditional "sage on a stage" lecture formats still cherished by faculty in many halls of higher education, but roundly rejected by growing numbers of students (Mazur 2009; Fitzpatrick 2012). First popularized by educational philosopher John Dewey in the early twentieth century, the engaged learning approach has gained increased attention in re-cent years for its capacity to move learning outside of the classroom and spark academic interest through interaction with "real-world" problems.

Some critics have questioned the novelty of these methods, noting that the active learning attributed to the "flipped classroom" model has been a fundamental operating procedure in humanities classrooms for decades (Davidson 2012; Brame 2013). Moreover, the excitement around recent models of "blended learning" seemed to some commentators to reflect techno-fetishistic attitudes toward shiny new gadgets rather than truly innovative characteristics of new pedagogical methods. Critiques of the tendency for education policies to promote investment in technology over investment in people led some weary observers to dismiss the wave of activity surrounding the rise of MOOCs and other models of online learning in the past decade. The collective FemTechNet developed an alternative model called the Distributed Open Collaborative Course (see Losh this volume). Our goal in creating the "Medicine in the Digital Age" MOOC was to prioritize the discussion forums as the site where the real learning would take place, and to leverage the global platform to engage a broader global community.

Phase Six: Medical Futures Lab in Historical Context

The debates about online learning reflect larger concerns about the role of technologically mediated experience in twenty-first-century life. In medicine, education, and beyond, we are grappling with our society's attraction to, dependence on, and vulnerability within the digitally intermediated world. In this context, two different, seemingly opposed imperatives have emerged for media scholars. First, the debate about supposedly new uses of screen-based media for teaching created an opening for historical dialogue about this century-long tradition, formerly called "visual education" when early twentieth-century teachers first started experimenting with using filmstrips and motion pictures in the classroom (Gaycken 2015). Almost 100 years ago, enthusiasts ranging from primary school teachers to insurance executives, public health officials to aviators believed that the emotionally immersive, rhetorically powerful approach to education through moving images would soon replace in-person teaching across content domains, age groups, and geographic locales. History has shown that the confident assertions of that era did not come to pass, in part because, like all technologies, the moving image technologies that promised to transform education in the 1920s were deeply entangled with the social contexts, political economies, and aesthetic traditions that surrounded them. The Progressive Era reformers found out, as others have since, that technology alone cannot change human behavior. As scholars who are trained to understand the complex interplay between technologies of representation and their effects on audiences, media scholars are perhaps uniquely well positioned to articulate a critique of the technological determinist enthusiasm around online learning. Discussion of the longer history of mediated learning can thus provide a grounded context for reflection on and intervention in contemporary developments.

On the other hand, the specific technological developments that created the conditions of possibility for the "flipped classroom," MOOCs, Khan Academy, and YouTube tutorials to emerge also created the opening for new kinds of applied media studies to emerge. Media scholars have shown how technological developments that bring innovative affordances together with specific historical and cultural contexts have led to the emergence of important new styles and philosophies of media. For example, Arthur (1993) demonstrated how the development of handheld, lightweight cameras and sound recording equipment in the late 1950s and early 1960s, coupled with the groundswell of civil rights and anti-war social movements that challenged dominant forms of political authority led to the emergence of the cinéma-vérité style of filmmaking. This style, characterized by wobbly camerawork, background noise, and seemingly spontaneous shot sequences became the signature anti-authoritarian visual aesthetic of the cultural revolutions of the 1960s. This example shows that when the right representational technology meets the right historical moment, it can have a transformational effect on cultural expression and contribute significantly to social change.

When we consider the technological transformations that made the online learning explosion possible, it becomes apparent that the same affordances of the mobile web that startups like Coursera, edX, Khan Academy, and others were deploying to personalize and democratize education were also being used by a wildly heterogeneous range of subcultures online. These groups found the video uploading, downloading, and streaming capacities of the social web to be ideal for information and communication sharing, also known as "hanging out, messing around and geeking out" online (Ito 2010). The key feature of these diverse forms of online content creation is that they are all primarily video-based. Whether synchronous or asynchronous, live-action or animated, long or short, sequential narrative or nonlinear, serial or one-off, purpose-built or passively captured, high or low production quality, high or low budget, firewalled or free, branded or anonymous, corporate or independently produced, the medium of communication is moving images with sound. Critically, these moving images with sound all live on the same internet and compete with each other for viewers' attention. After the launch of YouTube in 2005, it quickly became apparent that the competition would not only be fierce, it would also inspire participation by engaged viewers (Snickars and Vonderau 2010).

In other words, even though some critics have argued that most forms of online learning are more passive than—and therefore inferior to—in-person education, the broader trend of web-based learning is part of a larger movement toward highly engaged forms of audiovisual participatory culture (Jenkins et al. 2006). In addition to YouTube, platforms like Instagram, Facebook, Twitter, and other emerging social sharing sites are fundamentally driven by user-generated content. Without users who share video, photographs, links, and snippets of text, these sites would not exist. From this perspective, we can see that the technological affordances of the mobile, participatory web provided the elements necessary for applied media studies to come into being. These affordances did not cause any particular action or content to be produced, but rather, they created the conditions of possibility for actions and content to emerge.

While some of the Internet's trajectory was driven by an open, sharing, democratizing ethos, other forces pushed aggressively toward building an architecture of profit-driven, proprietary, closed systems on top of the participatory capacities of the web. The evolving field of digital health is shaped by the same tensions. At the present moment, medicine occupies a site on the digital health spectrum shared by highly resourced, expensive legacy electronic health record (EHR) systems and patent-protected drug and device development at one end, and do-it-yourself, peer-to-peer healthcare in online patient communities at the other end. To understand the intersection of the human and the technological in healthcare we must continue to explore both ends of the spectrum, through direct interventions that test the limits of critical-creative design. The next phase of Medical Futures Lab projects will work to bring critical perspectives to health technology startups through translational practices that craft health humanism as a high value return on investment.

Closing Words of Advice

In place of a conclusion, I have attempted to synthesize the lessons we have learned in the form of advice for different participants in the field of applied media studies. I hope they will be useful to future experimenters, creative hands-on critics, and their allies.

The building blocks:

1. Master multiple languages
2. Sketch out your vision in translational terms
3. Ask high-stakes questions that impact many diverse stakeholders
4. Invite lots of people to lunch
5. Find great collaborators who are different from you
6. Get institutional support (funds, physical space, public presence, leadership)
7. Don't waste time asking permission
8. Create environments that require/ensure regular output

The biggest challenges:

1. Sustainable funding model
2. Lack of training (how to be a PI)
3. Pipeline uncertainty (grad students, postdocs)
4. Competing institutional demands (time)
5. Uneven institutional support (across departments, divisions)
6. Different payscales (can be an asset in negotiations)
7. MOUs (avoid legal whenever possible)
8. Fatigue (heroic model)

Advice for individuals/scholars/experimenters:

1. Don't waste time arguing your relevance; show concrete examples
2. Keep your collaborators happy
3. Cultivate powerful allies
4. Follow university news and get to know the Office of Public Relations
5. Ask everyone for feedback and listen patiently.

Advice for institutions/administrators:

1. Support and reward creative faculty initiatives
2. Provide administrative and technical support for humanities labs
3. Support peer consulting on lab creation and sustainability
4. Don't expect your faculty to do all the fundraising by themselves
5. Provide teaching releases (this work takes a LOT of time)

References

Arthur, Paul. (1993) "Jargons of Authenticity (Three American Moments)," in Michael Renov (ed.), *Theorizing Documentary*, NY: Routledge, 108–34.

Balsamo, Anne. (2011) *Designing Culture: The Technological Imagination at Work*, Durham, NC: Duke University Press.

Baruch, Jay. (2017) "Doctors as Makers," *Academic Medicine* 92(1): 40–44.

boyd, danah. (2014) *It's Complicated: The Social Lives of Networked Teens*, New Haven, CT: Yale University Press.

Brame, Cynthia. (2013) "Flipping the Classroom," Vanderbilt University Center for Teaching. http://cft.vanderbilt.edu/guides-sub-pages/flipping-the-classroom/

Centers for Medicare and Medicaid Services. (2015) "Medicare and Medicaid EHR Incentive Program Basics." www.cms.gov/Regulations-and-Guidance/Legislation/EHRIncentivePrograms/Basics.html

Chan, Marie, Daniel Estève, Jean-Yves Fourniols, Christophe Escriba, and Eric Campo. (2012) "Smart Wearable Systems: Current Status and Future Challenges," *Artificial Intelligence in Medicine* 56: 137–156.

Chretien, Katherine, Neil Mehta, Warren Wiechmann, and Bryan Vartabedian. (2015) "Digital Literacy for Educators and Learners Toolkit," AAMC Website. www.aamc.org/members/gir/resources/359492/digitalliteracytoolkit.html

Crawford, Kate, and Meredith Whittaker. (2016) "AI Now: The Social & Economic Implications of Artificial Intelligence Technologies in the Near-Term," Workshop, New York, NY, July 7. http://artificialintelligencenow.com\

Davidson, Cathy. (2011) *Now You See It: How the Brain Science of Attention Will Transform the Way We Live, Work, and Learn*, NY: Viking.

———. (2012) "Why Flip the Classroom When We Can Make It Do Cartwheels?" *Fast Company*, May 9. www.fastcompany.com/1679807/why-flip-the-classroom-when-we-can-make-it-do-cartwheels

Dumit, Joseph. (2012) *Drugs for Life: How Pharmaceutical Companies Define Our Health*, Durham, NC: Duke University Press.

Fitzpatrick, Michael (2012). "Classroom Lectures Go Digital," *The New York Times*, June 24. www.nytimes.com/2012/06/25/us/25iht-educside25.html

Gaycken, Oliver. (2015) "Introduction on Displaying Knowledge: Intermedial Education," *Early Popular Visual Culture* 13(4): 249–55. doi:10.1080/17460654.2015.1130768

Greene, Jeremy A. (2016) "Do-It-Yourself Medical Devices: Technology and Empowerment in American Health Care," *NEJM* 374(4): 305–8.

Hagel, John, John Seely Brown, and Lang Davison. (2010) *The Power of Pull: How Small Moves, Smartly Made, Can Set Big Things in Motion*, New York: Basic Books.

IDEO. (2011) *Human Centered Design Toolkit*, 2nd ed. Ideo.org. http://www.designkit.org/

Ito, Mizuko et al. (2010) *Hanging Out, Messing Around and Geeking Out: Kids Living and Learning with New Media*, Cambridge, MA: MIT Press.

Jenkins, Henry, Katie Clinton, Ravi Purushotma, Alice J. Robison, and Margaret Weigel. (2006) "Confronting the Challenges of Participatory Culture: Media Education for the 21st Century." John D. and Katherine T. MacArthur Foundation, Chicago, IL.

Johnson, Steven. (2010) *Where Good Ideas Come From: The Natural History of Innovation*, New York: Riverhead/Penguin.

Kohn, Linda T, Janet Corrigan, and Molla S. Donaldson. (2000) *To Err Is Human: Building a Safer Health System*, Washington, DC: National Academy Press.

Kvedar, Joseph, Molly Joel Coye, and Wendy Everett. (2014) "Connected Health: A Review of Technologies and Strategies to Improve Patient Care with Telemedicine and Telehealth," *Health Affairs* 33(2): 194–99.

Losh, Elizabeth. (2017) "Rapid Response: DIY Curricula from FemTechNet to Crowd-Sourced Syllabi," in Kirsten Ostherr (ed.), *Applied Media Studies*, New York: Routledge.

Maeda, John. (2013) "Artists and Scientists: More Alike than Different." *Scientific American*, July 11. https://blogs.scientificamerican.com/guest-blog/artists-and-scientists-more-alike-than-different/

Mazur, Eric. (2009) "Farewell, Lecture?" *Science* 323: 50–51.

McPherson, Tara. (2009) "Media Studies and the Digital Humanities," *Cinema Journal* 48(2): 119–23.

Moss, Frank. (2011) *The Sorcerers and Their Apprentices: How the Digital Magicians of the MIT Media Lab Are Creating the Innovative Technologies That Will Transform Our Lives*, NY: Crown.

Ostherr, Kirsten. (2011) "The Medical Futures Initiative," unpublished pamphlet, October 17.

Ostherr, Kirsten. (2018) "Digital Medical Humanities and Design Thinking," in Olivia Banner, Nathan Carlin, and Thomas Cole (eds.), *Teaching Health Humanities*, New York: Oxford University Press.

Ostherr, Kirsten, and David Thompson. (2010) "Project TMC," unpublished pamphlet.

Prober, Charles G., and Chip Heath. (2012) "Lecture Halls without Lectures: A Proposal for Medical Education," *New England Journal of Medicine*, 366: 1657–59. www.nejm.org/doi/full/10.1056/NEJMp1202451

Rheingold, Howard. (2012) *Net Smart: How to Thrive Online*, Cambridge, MA: The MIT Press.

Snickars, Pelle, and Patrick Vonderau, eds. (2010) *The YouTube Reader*, New York: Columbia University Press.

Topol, Eric J. (2013) *The Creative Destruction of Medicine: How the Digital Revolution Will Create Better Health Care*, New York: Basic Books.

Verghese, Abraham. (2015) "Hope for Hands-on Medicine in the EMR Era," Medscape, April 6. http://abrahamverghese.com/wp-content/uploads/2015/07/2015-04-06-Hope-for-Hands-on-Medicine-in-the-EMR-Era.pdf

PART III
Challenges

6

PLEASURES AND PERILS OF HANDS-ON, COLLABORATIVE WORK

Kirsten Ostherr, Lisa Parks, Patrick Vonderau, Elizabeth Losh, Bo Reimer, Tara McPherson, Jason Farman, Heidi Rae Cooley, and Eric Hoyt

Kirsten Ostherr

In this collaboratively authored chapter, contributors discuss why they do applied media studies, addressing both the great rewards and the significant challenges to shifting away from traditional models of humanistic research and teaching. The guiding questions included,

> What has been your favorite part of doing applied media studies? What's your biggest headache? Why do you do this work? How has it impacted other things you do, like teaching, research, publication, dissemination, and service to the profession? Do you find it difficult to link your applied work to your teaching or research, or have these realms become synergistic for you? How does the time and energy required to cultivate and sustain applied media studies projects compare with the effort required by more traditional scholarly work? How does this kind of work fit into the annual academic schedules for course instruction and faculty evaluation?

This chapter also addresses the role of institutional setting, asking contributors to reflect on the ways that different contexts—such as urban research university, rural liberal arts college, elite private institution, or large, public institution—might shape the kinds of work possible, and the varying resources and constraints that these settings impose.

Notable recurring themes included the unpredictability of developing and integrating new forms of creative praxis into scholarly work, and the need to move outside of the physical spaces of the university. In doing so, many contributors note the intangible but meaningful benefits they gain from making their work "useful" to others, sometimes in unforeseeable ways, especially as participants from a broader public become involved. Often, contributors note that unexpected human benefits emerge from digital media entanglements, between actors who might not otherwise have had occasion to interact with one another.

What Has Been Your Favorite Part of Doing "Applied" Media Studies? What's Your Biggest Headache? Why Do You Do This Work?

Lisa Parks

My favorite part of doing applied media studies is engaging with and learning from people and communities beyond academic contexts and re-thinking media technologies and materialities from diverse perspectives. The biggest headache is trying to communicate across academic disciplines in collaborative projects that involve the design and development of tools based on fieldwork and community engagement. It is challenging to get computer scientists to fully appreciate humanities-based research methods and findings because our approaches are so different. Nevertheless, a headache can also become an exciting challenge. I think it is possible for media scholars to build bridges and collaborate more with computer scientists, electrical engineers, or architects. Given our expertise, it seems media scholars should participate directly in the design and creation of digital tools and media technologies.

Applied media studies implies a commitment to dynamism and creativity in academic work, and it allows for work in and out of academic contexts. Doing applied media studies also implies an interest in intervention and change, not just in the media sector, but also in society more generally. Some of the projects I have participated in have been related to underserved communities or political-economic issues such as globalization, surveillance, militarization, or environmentalism. Applied media studies helps to create a space in which to imagine the broader outcomes and effects of media scholarship and to find ways of sharing these outcomes with the various publics who fund and support our research and teaching.

Patrick Vonderau

My favorite part of doing this kind of work is that it feels like inventing your methods and objects from scratch. My biggest headache is that this feeling is always and ultimately deceptive. In other words, while applied media work comes with a feeling of freedom and invention, the downside is that such feeling easily leads us to de-contextualize and de-historicize what we do. On the one hand, there is an amazing array of new tools and discoveries they enable. On the other, there is a constant risk of giving up well-established standards— even such basic ones as compiling solid and exhaustive research bibliographies.

I do this work for various reasons: it's the difference between the kinds of work I have done that matters more to me than what these works may have in common. When it comes to our current work on Spotify (see: case study), there is just no other way to do this work. Method and object converge. Traditional media industry analysis, for instance, would have largely relied on interviews and trade papers to understand how Spotify partners with other actors in order to deliver its music streaming service. But you get to rather different insights— and a different idea of what Spotify actually is—when "sniffing out" the communication between Spotify and its business partners on the level of data.

Elizabeth Losh

I've enjoyed the many ways that being part of a large, distributed, global network like FemTechNet (FTN), which has over a thousand members all around the world, facilitates new contacts with designers, activists, and academics working in areas that I would never encounter if I were only siloed in my own disciplinary homes. It has also led to new kinds of collaborations and ways to connect and communicate with members of the general public. For example, because there was a group of researchers in FTN exploring how selfies might be important objects of study in digital culture, I became part of the development team for The Selfie Course (http://www.selfieresearchers.com/the-selfie-course/), an open access course about self-representation, ubiquitous computing, and social surveillance.

The big drawbacks have to do with the workload involved in participating in collectivity at such a large scale: trying to be all things to all people, particularly when traveling, commuting, changing institutions, or managing collaboration across several time zones. Mobility can be both a blessing and a curse when one is constantly connected.

I've always tried to write books that are intended to be interventions. It's important to me to address real-world problems with scholarship rather than retreat to an ivory tower. My first book was intended to help government agencies think critically about how tech culture propagates myths about political participation. My second book was intended to honor student-centered learning during a period of intense institutional investments in expensive and ineffective instructional technology. My co-authored textbook was intended to improve required undergraduate writing instruction by emphasizing student experiences that exist in specific contexts of identity work. The new book I've been working on with Beth Coleman about media activism is also structured around practical tips and lessons learned.

There aren't many other people at my institution working specifically on feminism and technology studies, so without FemTechNet collaborations I would guess that I would probably be a lot less productive. I feel I benefit from listening to and dialoguing with others and being part of a community that allows its members access to alternatives and a kind of "third place" that is neither a departmental nor professional workplace nor a domestic personal space (although it borrows from both). To work collaboratively with an emphasis on open pedagogy and social engagement, it has been wonderful to bring research, teaching, and service together in such meaningful ways.

Bo Reimer

There are two parts that I particularly like about doing applied media studies. The first is the—for me—necessarily collaborative aspect of it. Applied media studies is about doing things together, creating things—whether it be complex works for the Internet, digital tools, or physical installations in public spheres—that cannot be created by one researcher working in isolation. So it is about working with other researchers with other experiences and competencies, as well as working with actors outside the university, and in so doing both creating things that couldn't have been created in other ways, and as a side benefit learning new skills. The second is the experimental side of applied media studies. You try out things but you do not know where you will wind up. This unpredictability is also an important part.

The biggest headache is coordination. People have busy schedules and if you do not have specific funding covering everybody's participation, it is often difficult to find times for working together, no matter how exciting and important people think the collaborative work is.

For me this work is important in two ways. First, I do this kind of work because I believe it can aid in transforming situations and places. It is a kind of work which can have an immediate impact on a surrounding society. Thus, the interventionist part of applied media studies, as we do it, is crucial.

Second, doing applied media studies is also about changing the field of media studies; it is an intervention in how we do research and how we teach media studies. It is part of a general material turn within the social sciences and the humanities, but with some obvious exceptions, it is a turn that thus far has not had the impact on media studies that you would imagine it would have had.

Tara McPherson

I love the collaborative nature of how our lab does applied work. While humanities scholars often collaborate on projects like conferences or anthologies, our lab's work models a more far-ranging type of collaboration, a collaboration across and through difference. Our process brings together technologists, artists, scholars, librarians, and others to work deeply and iteratively together to create digital scholarship. This method of creating together is not always easy. Each participant brings unique talents to the table, but it can be challenging to find a common language as the collaboration unfolds. Learning side by side from others who have different methods and ways of thinking requires intellectual generosity and vulnerability. For instance, a literary scholar who has always produced print scholarship has carefully honed a particular way of writing. To be asked to think visually or to write to the structure of a database can be experienced as a deskilling. But we've seen amazing work emerge from this risk taking, and I very much enjoy building the environments in which this work can happen. I especially like the "code switching" that I do that helps foster communication across domains.

For well over a decade, I've had the privilege of collaborating with the same core design, editorial, and programming team, especially Steve Anderson, Craig Dietrich, and Erik Loyer, among many others. Our work together has been varied, from the digital scholarship published in the journal *Vectors* to the authoring platform Scalar.

My entry into practice-based work in digital media began from a deep curiosity about the ways in which communication was changing after the world wide web. When I first learned to create basic webpages with HTML or to create simple animations in Flash, I experienced the same giddy rush that I had felt editing video as a graduate student. So, at a very fundamental level, I do applied work because it feels different than the work I do with words. It takes me out of my head into a more intuitive relationship to my research materials. When Steve Anderson and I began imaging what *Vectors* might be, we wanted to tap into the expressive capacities of emerging screen languages. We wondered if you might play an argument like you play a game or be immersed in scholarship as you are immersed in a film. We were curious about the possibilities of interactivity for scholarship. In short, we were interested in how media scholars might not just theorize about media but also communicate in media in lively new ways. We also wanted to create spaces in which scholars working on issues we cared about—questions of affect, embodiment, gender, race, visuality, social justice, power, etc.—could explore the affordances of digital media for their own work.

So, we began out of a curiosity about what the digital might offer certain types of humanistic scholarship that print alone did not afford. As our work unfolded, we also became increasingly committed to rethinking scholarly publishing and to supporting open access scholarship and fair use. It's vitally important that we as scholars educate ourselves about the economics and politics of the publishing systems in which our work circulates.

Jason Farman

My favorite part of doing applied media studies is that it allows me to be a lifelong learner. I'm continually curious about how the technologies I use in my daily life work. My research focuses on mobile technologies, which are some of the most intimate devices that we use in everyday life. My mobile phone sits in my pocket throughout the day; it is the technology that is closest to my body as I move through daily life. It is my most intimate technology and I incorporate it in ways that allow the device to integrate into my life in a fairly seamless way. This has allowed the mobile device to be a part of my life in ways that often preclude analysis. So, my applied media studies approach to understanding mobile technologies has allowed me to ask new and interesting questions about these technologies, and to understand them in ways that most users of these media don't typically engage.

For my students, I see a similar kind of transformation as they engage in applied media studies with mobile technologies. These approaches allow us to defamiliarize the device. That is, we push the most familiar technology in our lives to become unfamiliar. This is very exciting to me because it allows us to think of these media in different ways. As I've incorporated these kinds of approaches into the classroom, I've been very clear that I am a co-learner with my students. I don't presume to have expertise in every area of mobile media. This allows me to learn new things, which gives me the opportunity to pursue my curiosity without needing to master things before bringing them into the classroom. And second, it allows me to encourage my students to rid themselves of the notion of mastery. Many of my students feel that if mastery has not been attained, then they have either failed, or what they have studied has not been of value. Removing the constraints imposed by notions of mastery gives us the freedom of exploration and experimentation that are the foundations of applied media studies. Ultimately, the things I find most thrilling about applied media studies are its ability to encourage curiosity, experimentation, and the exploration of new pathways into very familiar media.

I believe that making is a mode of thinking. I believe that there are certain ways of knowing that require an applied media studies approach. To be explicit: I am a strong advocate for the kinds of theoretical engagements that give us the tools for certain modes of abstract critical thinking. The work that I did in my first book, *Mobile Interface Theory*, was grounded in theories of phenomenology and poststructuralism, and I use these theoretical approaches to analyze and understand how we use mobile technology. Out of this, I created a theory of the body that I termed the "sensory-inscribed body." As my work in this field continued to develop, I began to understand the sensory-inscribed body in a different way, one that better incorporated the ways that everyday objects shape the formation of this body. The ways that we live and practice our bodies are deeply shaped by the objects we encounter. Applied media studies is one way to think about the relationship between the body and the objects that we encounter in everyday life. As we become critical makers, the connection between our bodies and technological objects in the world is made explicit. So, while critical theory has allowed me the avenues to understand the body and to begin to create a new theory of the body, applied media studies and hands-on explorations have given me new methods to answer the new questions that emerge out of critical making.

Heidi Rae Cooley

What I like most about doing "applied" media studies is that the "doing" is collaborative. Thinking and working with colleagues and students from a variety of disciplines is generative of more productive or progressive understandings of, for example, what constitutes an effective and socially significant interactive media experience. That these conversations also include Ward One Organization members and other community stakeholders means that the work of project development attends to the concerns of a broader public: we do not collaborate in a silo. This, of course, is also a principle challenge. Multiple voices necessarily complicate decision making and can interfere with deadlines. Intensive discussion and active listening become fundamental to how design goals are imagined and implemented, and how impactful content is produced. The ongoing negotiation across processes of conceptualization, implementation, and testing—especially as these processes are experienced and interpreted with others iteratively over time—has proved generative of new and shared appreciation for, in this case, history and how history shapes the contemporary landscape of Columbia, SC. Despite the occasional messiness of the work, it is immensely rewarding.

I am committed to this kind of work because it produces meaningful relationships across disciplines and more progressive and provocative understandings of place and context beyond the university. The project I'm focusing on here makes visible otherwise overlooked or ignored histories of racial inequity. Collaborating with former Ward One residents to present the history of the old ward and their memories of and reflections about it has helped to begin a mending of divides between the University of South Carolina and former Ward One residents. Our work provides a space for cultural renewal. Students, faculty, and the community all benefit as they begin to appreciate one another and each person's contributions to civic life.

Eric Hoyt

The best part of applied media studies is hearing from users. I love it when people tell me that Lantern transformed their research process, helped them better understand a relative, enabled them to stay put with their family rather than travel to a library across the world, anything that enhanced their life in some way. The biggest headache comes when users point out Lantern's many metadata errors, missing images, pagination glitches, etc. I'm grateful to have these called to my attention, but I'm usually too busy teaching, writing, and parenting to fix these things as quickly as I would like. The headache is the gulf between the user-experience we want and where it stands, and the limited resources that we have most of the time to address it.

I do this work because I see the quantitative and qualitative impact. According to the Internet Archive's metrics, items from the MHDL's collection have been downloaded 3,893,433 times. From its public launch in the summer 2013 through May 2016, Lantern has attracted 133,024 unique users, with the average user spending 11 minutes 14 seconds per visit. I'm spending time this week working on Lantern again because I want to do right by all these people, even if I don't need most of the artifacts that I'm editing and indexing for my own research.

How Has This Work Impacted Other Things You Do, Like Teaching, Research, Publication, Dissemination, and Service to the Profession?

Lisa Parks

In teaching, this work has led to a plethora of hands-on or applied exercises and assignments, ranging from drawings to site visits, from object analyses to interviews. In the class I am teaching now I asked a graphic illustrator to come in and work with students to generate drawings related to a writing assignment they are working on. In other classes, I have taken students to sites that are relevant to our course and asked them for a collective analysis of it. I have used some of these exercises in my own research too, whether to catalyze a new line of inquiry, think about relations between disparate things I want to write about, or develop new concepts. Working in this way across teaching and research has become synergistic for me.

Patrick Vonderau

I greatly admire colleagues who have managed to establish synergistic relations between their applied research, teaching, and service. My own work, however, is split up between what happens institutionally, within the confines of my department, and outside the institution: via mail, personal encounters, at conferences, during research interviews, or occasionally in Umeå University's Digital Humanities lab where our Spotify project is situated. I do cinema studies teaching at my department and digital projects elsewhere. And I am not convinced that my cinema studies' work necessarily benefits from the kind of Big Data analysis that is now often seen as identical with digital humanities (DH) projects. While my department certainly appreciates this other kind of work, fitting it into the curricular needs of what and how we teach cinema studies feels a bit artificial. There is not a total or ideological divide here, but a division of labor which I feel is entirely justified.

Elizabeth Losh

The work may be satisfying, but many structural problems exist to having it be rewarded. Social engagement gets coded as low-status "service." Collectives don't generate single authored monographs; they write manifestos. And digital production continues to be devalued.

I benefit from a position of obvious privilege—as a tenured professor with a full-time job. So doing co-authored or collectively authored works doesn't harm my career, and the cross-campus labor is legible to my institution as "digital humanities" because I have a track record of publication in that area (even if I am trying to push the field by looking at phenomena like hashtag activism and talking about issues like appropriation when it comes to data ethics). Others in the network contribute a lot more labor with a lot less reward, so figuring out how to signal boost for people doing good work on feminism and technology is important, particularly because many of them are in more precarious conditions.

It can also be a challenge for students to engage with a globally distributed network. It can sound exciting for a class to be connected with a group of activists working on street harassment in Bangalore, but students need context, connection, intimacy, and meaning-making. You can't just Skype in random strangers and assume that there will be a meaningful educational experience, even if your own research might seem exciting to you.

Bo Reimer

Having worked for fifteen years at a school that was founded on the notion of the integration of theory and practice, and having taken part in the setting up the school, my research and the courses I teach are basically always carried out through the perspective of integrating theoretical/critical work with practice based work. That meeting point is in many ways the point of departure for the work done both by me and my colleagues.

Tara McPherson

In many ways, the work we do at the Vectors lab emerged at least partially out of experiments in teaching. Soon after arriving at the University of Southern California (USC) as a young professor, I was recruited to teach for the Institute for Multimedia Literacy (IML), a space that helped faculty integrate multimedia authoring into the classroom. This experience at IML helped fuel my interest in digital scholarship and led to discussions with my dean about launching the journal. Many of my courses now incorporate practice components, and the IML has expanded to become the newest division of USC's School of Cinematic Arts, home to our practice-based Ph.D. program.

My applied work has also led me to develop relationships with scholarly societies, foundations, and agencies. As our lab has moved from developing pieces for *Vectors* to creating authoring software, I'm involved much more deeply in infrastructural questions around scholarly publishing. I now often find myself in conversation with librarians, archivists, funders, presses, and societies on an ongoing basis. I have a better sense of the larger ecosystem that supports scholarly communication and am deeply engaged in this work through a variety of national task forces and committees. It's not the sexiest part of what our lab undertakes, but it's very important to understanding larger scholarly systems.

Jason Farman

There is a significant synergy between critical making and the modes of thinking that are articulated in my research. The critical making that informs my research largely comes from my work with students in the classroom. I find it important to practice applied media studies with people from vastly different backgrounds, cultures, and skill sets. In my particular embodiment, I engage in applied media studies in a very specific way based on personal histories and contexts. Thus, doing applied media studies as a single practitioner within my personally specific contexts is ultimately limited. Applied media studies works best when many different bodies are engaging a single object or a single media context. Each person brings their histories and context to a media object that allows others a different perspective on that medium. Here, I appreciate Sara Ahmed's work in *Queer Phenomenology* (2006), where she writes, "An object tends toward some bodies more than others… So it is not simply that some bodies and tools *happen* to generate specific actions. Objects, as well as spaces, are made for some kinds of bodies more than others" (51).

The classroom is an important place to see this kind of work in action. For example, as my students begin to do usability testing of mobile apps, many of their assumptions about how people will encounter those apps are laid bare. The notion of an "ideal user" of a particular medium or the idea of a universal body that will encounter something like a mobile app becomes deeply problematic through an applied media studies approach. As a researcher, I find the classroom to be a rich site for understanding the ways that various bodies encounter these different media. My scholarly writing and analyses would be significantly lacking if it weren't for the important ways that my students teach me about the range of contexts and approaches to these media.

Heidi Rae Cooley

I am fortunate. The applied work I do is integral to my other professional activities and to how I pursue my research. For example, the Ward One project is an ongoing pedagogical exercise; my collaborator Duncan Buell and I team-teach a cross-College course, Critical Interactives, once each academic year. At the same time, the project offers a unique opportunity for me to think *through* theoretical questions that underpin my scholarship, specifically questions about habit-change in the mobile connected present. And it has served as a case study for conference papers and single- and co-authored articles.

How Does the Time and Energy Required to Cultivate and Sustain Applied Media Studies Projects Compare with the Effort Required by More Traditional Scholarly Work?

Lisa Parks

This is an intriguing question and I was thinking about this the other day, based on how much time I was spending to create visual materials related to a lecture that I was giving. I not only had to write a full paper but then to create an entire sequence of layered images, which were designed to advance my argument visually. Writing is the core of what I do, and I love writing. Yet I am always writing in parallel with other activities, whether fieldwork, media consumption, mapping and visualization, art collaboration, digital tool generation, and so on. It is challenging to do these things simultaneously and sometimes feels overwhelming, but it also keeps work unpredictable and open. Because we live in such a logo-centric society, there is little space in academia for synesthetic modes of academic production. Scholars have been doing this type of work for decades—it is not just a phenomenon of the digital age. Yet the myriad forms of creative labor that scholars perform alongside their writing often goes unnoticed and tends to be undervalued.

Patrick Vonderau

I do not think we need to over emphasize or essentialize the difference between traditional humanities and applied media work here. Both take a lot of time and energy. It's not a new, just a different kind of work. I have spent years in archives and years elsewhere. The second question is, again, a question related to the institutional aspects of this form of scholarly engagement. I am privileged to work at a progressive Swedish university that sees innovative work as a merit, and is ready to acknowledge the value of such work. In other words, digital initiatives, experimental curriculum development and other unorthodox activities are very much appreciated.

Elizabeth Losh

The biggest problem is that applied media studies often requires being off campus, and that's true whether you are looking at communities next door or on the opposite side of the planet. I used part of my sabbatical to visit four teams of researchers working on mobile technology and financial inclusion who were doing fieldwork with different groups in four different regions in India: rickshaw pullers in Delhi, fisherman in Kerala, rural women in Bihar, and silk workers in Karnataka. During the academic year that trip would have been impossible without disrupting my teaching, and during the summer researchers aren't *in situ*.

Bo Reimer

The one thing that distinguishes applied media studies projects from other research projects is the need for long-term commitments. These are projects you have to invest lots of time on. You have to build up trust with your co-partners and think through very carefully before you start a project whether you actually will have the time and energy necessary to carry it through. You also have to recognize that there may be no obvious end dates for such projects, and that different partners will have different views on when to terminate the projects.

Tara McPherson

The number of words I've written for grant proposals and committee reports is probably on par with the books and articles I've authored. Our lab largely runs on external grants, and these grants require a great deal of administrative labor. Very little of the training we receive as humanities graduate students or faculty prepares us for this type of management or to wrangle Excel spreadsheets. It's certainly not my favorite part of my "applied" work, but it is crucial to our efforts. I also spend many hours a month sustaining relationships with our various partners, from academic presses to archives and humanities centers. We also run workshops and help others begin to write grants.

None of this labor is easily translated into the university's metrics for faculty evaluation. Tenure and promotion committees in the humanities are flummoxed by non-traditional projects and are often ill prepared to evaluate digital scholarship, collaborative research, and grant-funded projects. There are signs of change in this regard, as scholarly societies and foundations have helped fuel larger conversations. I recently served on the College Art Association's task force on evaluating digital scholarship and am glad such work is underway more broadly. There are also key questions about fair labor practices that need to be addressed. If the work of 'making' is often undervalued by tenure committees, those who collaborate on making projects without the security of tenure-stream jobs are much more vulnerable than faculty. We need better ways of acknowledging and rewarding every team member.

As I read the answer I've provided here, I take note of words like "management" and "spreadsheets." I sometimes worry about the unintended consequences of the work that we do. It's not inconceivable that the turn to "applied" scholarship or teaching could hasten the managerial turn already underway in the university system. Many critics of the digital humanities and of online learning have made this point. It's important that in privileging the applied elements of media studies that we not neglect the intellectual, political, and theoretical dimensions of the field. Again, we need to hold theory and practice in productive tension and not get lost in our spreadsheets.

Jason Farman

Personally, I found that the time required for doing the work of critical theory for my research was much shorter than the time required to do applied media studies. Working with people often takes longer, and the approaches to applied media studies that I advocate require researchers and students to engage as many different people from as many different backgrounds as possible. This takes time. So, while the output for this kind of scholarly work might be slower, I have found that it creates dynamic teaching environments that have been positively evaluated and rewarded in traditional academic settings. At Tier-1 research institutions, academic publications will always take priority over dynamic teaching; personally, I have found that the kinds of work that are valued for tenure and promotion aren't necessarily in line with the reasons I decided to pursue this career in the first place. For me, creating dynamic teaching environments and transforming the ways that my students think about their technologies will always take precedence over creating substantial academic output in the form of scholarly publications.

Heidi Rae Cooley

For me, the two approaches to scholarship are mutually informing. As I indicate above, my intellectual efforts transpire in tandem with project development in very concrete ways. That said, applied media studies, as I understand it, requires a particular kind of time and patience. This is because it requires cultivating relationships and building trust both institutionally and with members of the community. And it doesn't confine itself to an academic calendar. While limited resources mean that project development slows, for example, when my collaborator and I are not team-teaching the Critical Interactives course, we are actively seeking funding opportunities, reaching out to individuals and organizations who might provide assistance or insight, and maintaining an ongoing conversation with former Ward One residents.

Eric Hoyt

As I discuss more in my chapter about Lantern, the labor required to sustain a digital project marks a huge difference between applied media studies and traditional scholarly writing and publishing. On my CV, everything gets reduced to a single bibliographic entry, generally with a single date. It's difficult to communicate and appreciate that a web project that took hundreds of hours to develop and launched in 2013 has also taken hundreds more hours to sustain and update ever since.

What Role Does Your Institutional Setting Play in Shaping the Kinds of Work You Find Possible to Create?

Patrick Vonderau

This question implies that institutions can be a hindrance, and that tells us a lot about how "applied media" work has developed out of and beyond the borders of institutionally defined "studies." I would suggest making a difference here between our local teaching and research environments (our departments or universities) and the institution of academia in a more abstract sense, as constituted by set scholarly standards and best practices, keywords and paradigms, or traditions of method and theory. The departmental setting I work in is only partly overlapping with the institution of media studies. Stockholm University is Sweden's most important research university, and its Department for Media Studies is the largest and most diverse in Scandinavia. While this opens up many possibilities for exchange, cooperation, and experimentation, there are of course traditions and strands of research that shape our conversations. And sometimes you need to talk to people elsewhere in order to develop new ideas.

Elizabeth Losh

Moving from a more internationalized and somewhat more urban research university to a campus that conceptualizes itself more as a liberal arts college could pose some interesting challenges if you are interested in applied media studies (although both are state public institutions governed by similar bureaucratic rules with strong national rankings that can contribute to entitlement and elitism). Fortunately my new campus has a strong commitment to experiential learning, information literacy, and global awareness, so my colleagues' attitudes are hardly provincial despite our location.

I also think regional collaboration is very important for building scholarly friendship ties. Collaborating over greater distances can also contribute to more breakdowns of communication and legitimately hurt feelings and microaggressions. I benefitted from being part of a strong FemTechNet cohort in Southern California, and now I am building ties in the DC/Virginia/Maryland region by developing the Equality Lab project.

Figuring out how to be a better colleague is critical to avoiding resentments. I am trying to work more on listening, mentorship, not delegating unpleasant work, and avoiding the bad habits of executive crisis management I picked up as an administrator in very large courses with many stakeholders. It's an area where it is important to apply one's feminist values and be attentive to the needs of community members, curators, archivists, librarians, IT support personnel, and lots of people who do the work of care and repair in applied media studies.

Bo Reimer

I have been lucky in the sense that my university prioritizes action-oriented work, and work involving the community we belong to. The research center that I am director of has a base funding from the university. The funding is not huge, but it is a resource that covers a studio and some funding for the work to be carried out.

Tara McPherson

USC is a large, private, urban university that has historically had strong professional schools. I'm sure my career has taken different turns because I work within a cinema school in the shadow of Hollywood rather than in an English department in Cambridge. I teach side-by-side with screenwriters, documentary filmmakers, film historians, production designers, feature directors, video game designers, media theorists, and feminist experimental animators. This setting has given me the freedom to take risks and to embrace applied work, even if the larger university still struggles to make sense of practice- or arts-based research.

USC is also an overtly entrepreneurial university. This has both advantages and drawbacks. On the one hand, the university is not especially beholden to tradition and is open to new ideas. On the other hand, it doesn't always have in place the structures needed to support its new ideas. When I began writing grants within the cinema school, the business office had little experience in this regard. We had a lot to learn and made mistakes along the way.

Jason Farman

Having worked in a range of academic settings, I find it beneficial to be in an environment that fosters diversity and inclusion. Applied media studies as an approach functions best when there are many different perspectives brought together. Not all institutional environments prioritize diversity and inclusion. In fact, many of the foundational structures of higher education limit the kinds of diversity that we can draw on for the work that we do. Right now, I am lucky to work in a program that is separate from my students' majors. Instead, my program (called Design Cultures & Creativity) draws on students from as many as forty-five different majors across the campus. The advantage of this approach is that we bring students with many different academic contexts and backgrounds together and get them to collaborate in ways that are difficult within a traditional university major. We are also a living-learning program, so our students live together in the same building in which we have our classes and our lab space. The result is that we bring together engineers, computer scientists, Women's Studies majors, artists, business majors, and architects to approach and explore a single problem in a hands-on way from many perspectives. Our students come to us not only from a range of racial and ethnic backgrounds, but also from a range of disciplinary backgrounds. To me, this is an ideal setting for the kinds of applied media studies that I hope to do. It extends beyond traditional disciplinary boundaries to engage media in a way that values our diverse perspectives and experiences.

Heidi Rae Cooley

The University of South Carolina is the state's flagship research institution. While resources are not abundant, there is interest in and support for interdisciplinary endeavors. For example, the university's Center for Digital Humanities (CDH) has been a site for conversations about project development and exhibition/dissemination, and our collaboration is recognized as a model for this kind of work on campus. That I'm jointly appointed in the School of Visual Art and Design in Media Arts and the Film and Media Studies Program has also facilitated my collaborative work. Both department and program are supportive of the Critical Interactives class, and those who oversee class scheduling for each unit cooperate with the Department of Computer Science to ensure that we can offer the course each year. There are internal grant opportunities that we have benefitted from. And curators for the Moving Image Research Collections and Digital Collections, along with the University Archivist and Oral Historian, have all been instrumental to the development of our projects—conducting workshops and offering tours, assisting students with accessing various audiovisual materials, attending demonstrations, and offering feedback. Moreover, I am extremely fortunate to have generous colleagues who have contributed to the development of our projects by raising questions and offering constructive criticism.

Reference

Ahmed, Sara. (2006) *Queer Phenomenology: Orientations, Objects, Others*, Durham, NC: Duke University Press.

7

MEDIA FIELDWORK

Critical Reflections on Collaborative ICT Research in Rural Zambia

Lisa Parks, Lindsay Palmer, and Daniel Grinberg

This chapter explores what it means to conduct media fieldwork by critically reflecting upon a collaborative research project in rural Zambia in 2012 and 2013. It discusses what a media studies approach to fieldwork entails and what the affordances and limits of such an approach might be. By fieldwork, we are referring to the practice of traveling to a location, visiting various sites in and around that locale, interviewing and conversing with people in structured and semi-structured ways, and observing any number of activities, objects, or relations that help to characterize or define media-related phenomena at the site. Fieldwork also encompasses the complex issues and relations that emerge in the context of such investigations. This method should account for the political and ethical dimensions of conducting site-specific research in others' communities. For one, it should recognize the ways that legacies of colonialism and conflict, as well as intersectional differences can inform or shape research interactions and processes. Fieldwork also involves considering questions of labor as well as determining how collaborative work is conducted. Fieldwork, of course, has a long tradition in anthropology, sociology, and other fields, and comes with specific standards and expectations for those disciplines.

The fieldwork practices adopted in this project were influenced by research in the fields of visual anthropology and information and communication technologies for development (ICT4D). These fields often use methods of ethnography, site visits, and community engagement. For instance, scholars such as Jenna Burrell (2012), Brian Larkin (2008), and Debra Spitulnik (1998, 2002) have generated valuable ethnographic studies of contemporary media cultures in Ghana, Nigeria, and Zambia, respectively. While there are significant overlaps between the fields of media studies, visual anthropology, and ICT4D, it is important to note that the academic training, reading, research questions, and methods or critical practices of humanities-based media scholars are distinct from those of anthropologists and ICT4D researchers. Media scholars with advanced degrees typically acquire expertise in the political economy and state regulation of media institutions and corporations. They study the international history of modern media technologies and cultures, sometimes called "media archaeology" (Huhtamo and Parikka 2011). They learn about the globalization of media industries, as well as the semiotics and formal properties of audiovisual representations that appear on multiple kinds of screens and interfaces. They also engage with particular lineages of philosophy and cultural theory, ranging from formalism to feminist criticism, from post-structuralism to theories of globalization. In general, the goal of critical media scholarship is to understand the particularities and dynamism of media institutions, technologies, formats, and cultures, rather than to offer an immersive account of a social structure or cultural milieu.

While anthropologists' and ICT4D researchers' ethnographic studies of modern media cultures in Africa have informed our research, our fieldwork does not aspire to or adhere to all of the central tenets of a classically defined ethnography: immersion in spaces, mastery of languages, establishment of cultural competency, longitudinal study, participant observation, and "thick description." As media scholars, we are primarily interested in understanding how diverse communities in the world think about, organize, and use media technologies to support their interests. To investigate such issues, we employ ethnographically inspired research practices, including fieldwork. In this way, our work builds upon the site-specific research by media scholars that has emerged over the past decade. The collection *MediaSpace: Place, Space, and Culture in a Media Age* (Couldry and McCarthy 2004) set an agenda to situate the discipline's studies within a larger constellation of spatial dynamics. Nick Couldry and Anna McCarthy observed that "a geographically informed and spatially sensitive analysis of media artefacts, discourses, and practices reveal forms of inequality and dominance, knowledge and practice that are hidden from other analytical techniques" (4). Other works have focused on topics such as the various sites of media reception, from the airport to the waiting room (McCarthy 2001), the studio milieus in which media are made (Caldwell 2008), the assembly lines on which TV sets are manufactured (Mayer 2011), and the infrastructure sites through which media signals are transmitted (Parks and Starosielski 2015; Starosielski 2015). Such studies often privilege the urban over the rural. As Couldry remarked in *Media Rituals* (2003), there has been a tendency in media studies to focus on media centers and cities rather than media peripheries. This has resulted in less site-specific research in rural areas by media scholars, though there are important exceptions such as Eric Michaels' (1993) research in central Australia or Darin Barney's (2011) research on the Canadian prairies.

In an era of intensifying media globalization and digitization, it is crucial that media scholars experiment with and create methods and critical practices that can expand understanding of the ways media technologies and content are organized, regulated, and used by different people in different parts of the world, including in rural communities. In his provocative book *Whose Digital Village?* (2017), Ramesh Srinivasan insists that the democratization of media technologies involves much more than ensuring access or connectivity: it requires prioritizing concepts of difference, ontology and voice in design processes and shifting from an industrial paradigm of world making to one of "world listening." Other media scholars have underscored the importance of global/local and regional approaches. Rather than idly accept media globalization as Hollywood's *fait accompli*, they have examined the dynamic operations of media industries in areas like China, the Middle East, Eastern Europe, Latin America, and India, (Curtin 2007; Imre 2009; Kraidy and Khalil 2009; Venegas 2010; Govil 2015) and explained how local, regional, and national forces battle and negotiate economic and cultural imperialism in the media sector.

The goal of media fieldwork is not to produce a masterful account of Others' media cultures; rather, it is to generate a space of transcultural interaction and exchange through which multiple knowledges about media can take shape, among and between researchers and members of the communities in which research occurs. Because fieldwork inherits the baggage of history, including (post)colonial tensions, the terms of these interactions and exchanges are never easy or egalitarian. They can be fraught and challenging. Such conditions should not inhibit media scholars from trying to learn from and understand how diverse people beyond the world's media capitals participate in media systems, economies, and cultures. Ultimately, our interest in fieldwork coincides with an interest in understanding how material conditions, location, *difference*, and *power hierarchies* function as part of media cultures. It is our belief that media scholars have not yet recognized or learned from an array of media-related organizations, sites, objects, users, and experiences that exist on the outskirts or fringes of media capitals. To provide further detail about our approach to applied media studies, we undertake a critical analysis of media fieldwork we conducted in the rural community of Macha, Zambia (Figure 7.1). This account will highlight the challenges and complexities of doing such work, as well as its epistemological affordances and limits. We conclude with a brief discussion of four principles that are important to consider when conducting future media fieldwork in developing rural communities.

Fieldwork in Macha

Macha is a rural community located in Zambia's Southern Province, 70 kilometers away from the nearest town of Choma and almost 350 kilometers from Lukasa, Zambia's capital. Macha is governed by a local Tonga chiefdom and a Brethren Church in Christ mission, established in 1906, is situated in the community's center. The mission complex includes several schools, a regional

FIGURE 7.1 A view of the dirt road into Macha's center, which people use to move through the area by foot, bicycle, car, and mini-bus each day.

Source: Photo by Lisa Parks.

hospital, a malaria research institute, a police station, a community center, a barbershop, housing, and an open-air market. Most Machans live in rural homesteads without electricity or running water and work as subsistence farmers who grow corn, cotton, peanuts, soybeans, or sunflowers. Many people in Macha speak English and most speak Chitonga. In 2009, an organization called LinkNet, supported by a Dutch nonprofit organization, established internet service in the community using a VSAT connection and repurposing old cargo containers as internet access stations. LinkNet became part of a broader community initiative called Macha Works, led by Dutch engineer Gertjan van Stam and a team of Machans (van Stam 2011, 2014; Mudenda and van Stam 2012). The initiative emphasized local empowerment and established an information technology school, new apartment housing, a radio station, a restaurant, and sustainable farming programs. However, due to a series of community conflicts, LinkNet ceased service in 2012 and Macha Works was shut down as well. These developments caused disappointment and strife within the community. Around the same time, two Zambian commercial mobile service providers, MTN and Zamtel established service near the center of Macha. Machans could access the internet by purchasing data packages or "talk time" from their mobile phone providers or by purchasing a portable modem or "data dongle" for use with a computer. Yet the costs remained prohibitive for most people in the community (Pekovic et al. 2012).

Members of our research team visited Macha for a total of six weeks in 2012 and 2013. We were invited to participate in the project by computer scientists at UC Santa Barbara,[1] who had already been collaborating with LinkNet and Macha Works on traffic analyses to assess local network efficiency. These computer scientists sought to develop the project further using interdisciplinary collaboration, community engagement, and technical deployments, and asked Lisa Parks to serve as a principal investigator in a grant proposal to the National Science Foundation, and lead the fieldwork. The overall goal of our research project was to work with the community to develop technical solutions that would support local control over networked information sharing and digital content generation. These technical solutions were to be "socially informed," meaning that they should be shaped in relation to input from community members. As media researchers on the team, we used fieldwork practices such as community engagement and collaboration, interviews, and site visits. Our goal was to gather input from people in Macha and study existing network conditions and content generation in the community (Parks 2015, 2016). We shared our findings with the computer scientists on our team who developed a Facebook app called Village Share and a free local mobile phone service called Village Cell (Johnson et al. 2012; Zheleva et al. 2013). In what follows, we describe and assess our media fieldwork practices and discuss the advantages and shortcomings of our methods. This account is a means of flagging the complexities associated with media fieldwork and sharing insights with others who may be undertaking similar studies.

Community Engagement and Collaboration

When we arrived in Macha, one of the first things we did was to meet with our local partner, Consider Mudenda, Director of the Linknet Information Technology Academy (LITA), to have a detailed discussion of the research project. Months before our arrival, we had initiated regular online communication with Mudenda to discuss the project and solicit community participation, which he was organizing. Being able to meet with Mudenda and his team of six local collaborators, all of whom spoke English, was vital. Meeting in person set the tone for the research activities that followed and provided an opportunity for all members of our team—visiting researchers and local partners—to openly communicate their needs and expectations about the project. We also established a plan and schedule for the collaboration and discussed a number of issues at our preliminary meeting. The topics included an introduction of all participants involved, overall goals of the project, the terms of the collaboration, the research methods used, sites to be visited, translation and transcription practices, logistics and transportation, gaining of interviewees' consent, community concerns about the project or the issues it explores, and anticipation of research output (e.g., research presentations and publications, technological solutions, blog posts). This preliminary meeting helped to emphasize the professional standards of the research, while personalizing the process and ensuring that participants were able to express and resolve any concerns at any time throughout the research.

To encourage collaborators' input we held brief meetings at the end of each day, so that team members could provide status updates and feedback on the research process. We found that our partners were a bit shy and reluctant to speak up at first. However, as we got to know one another a bit better, our collaborators became more forthcoming and offered more thoughts about the project. Sometimes their comments were brief, and sometimes they were longer and more detailed. We noticed that there was intense interaction amongst the local partners at the end of each workday. They also tended to discuss the project and what they gained from it amongst themselves instead of directly with us. Our efforts at community engagement—that is, involving Machans directly in our research as partners and participants—may have generated more intensive internal discussion among community members than between the community and our group. There was some reticence in discussing the research issues and questions in depth with us, which may be related to the limited history of ICTs and digital media in the community, perceived cultural/language differences and knowledge hierarchies, uncertainties about the collaboration (local partners kept a distance while still working with and assisting us, which suggested a simultaneous reticence and curiosity), and generational differences (our local partners were younger than our research team members and may have been trying to be respectful or deferent to us). Despite this, our collaboration with local partners allowed us to engage with a broad cross-section of perspectives in Macha. At the end of the project, our partners also informed us that they were grateful to have the chance to talk to and hear from people in the community about the ICT issues we were interested in investigating.

Interviews

One of the primary methods we utilized as part of our fieldwork was the interview. Working with our local partners, we recorded video interviews with more than 200 community members to better understand how Machans think about and use information and communication technologies in their everyday lives. Before arriving in the community, we created a list of questions and shared these questions with Consider Mudenda via the Internet. We solicited feedback and comments from him, and revised the questions accordingly. Once in Macha we read through the questions again with our local partners and did another round of revision to eliminate technical jargon, streamline language, and clarify terms. In an effort to learn from as many people in the community as possible, we adopted a dual approach to the interviews. First, Mudenda organized interviews for us with people who might be described as Macha's ICT early adopters and influencers. This group included internet administrators, chiefs, teachers, radio DJs, hospital staff, church workers, farmers, and students. We conducted these interviews, and refer to them as *interviews by visiting researchers*. Second, working with Mudenda in both 2012 and 2013 we appointed a team of six local collaborators.[2] We asked them to partner with us and use our questions as a basis for video recorded interviews with Machans throughout different parts of the community. Each team consisted of three women and three men in their late teens or early twenties. Most of these partners had taken the ICDL (International Computer Driver's License) training course from Mudenda at LITA. Thus, they had basic proficiency with computing and web navigation, though none had regular internet access. We refer to these conversations as *interviews by community partners*, because they were conducted by people in Macha who collaborated in our research.

Interviews by Visiting Researchers

Working with Consider Mudenda, we conducted videotaped interviews with individuals and focus groups in Macha and neighboring villages. Usually these conversations were in English, but sometimes we worked with a translator. These interviews were usually scheduled in advance and lasted anywhere from thirty to ninety minutes. We conducted these interviews in local schools, in peoples' homes, at the hospital, community centers, public market, fire camps outside the hospital, and in nearby communities such as Mapanza and Chikanta. We also conducted interviews with Chiefs Macha, Chikanta, and Mapanza at their palaces. Our partner Mudenda often facilitated these meetings by introducing us, helping with translation when needed, and providing transportation. At times Mudenda also participated in the interview process, asking his own questions and discussing the results of our interviews with us. We began with an interview protocol but found that converting it into a more semi-structured and conversational format worked better. The latter approach put the interviewee at greater ease as we asked about his or her ICT knowledge and usage. The formality and length of these interviews varied depending on the informant's background, the location, and the flow of discussion. For instance, those who worked in hospitals or at the schools and were socialized into the mission's culture often expected a level of formality that was not expected by farmers who were staying at the fire camps outside the hospital.

We attempted to engage with Machans who held a variety of social positions to gain different perspectives on the meanings and uses of ICTs in the community. For instance, to understand the perspectives of community leaders, we interviewed Chief Macha at his local palace. This interview was video recorded with his permission and lasted about ninety minutes. After the interview, Chief

FIGURE 7.2 A view of the fire camp area adjacent to Macha's regional hospital, where people wait for visiting hours and cook for family members and friends.

Source: Photo by Lisa Parks.

Macha took us on a tour of his home, which included a spacious garden and pig farm. He then proudly showed us the latrine he had built as a model for those in his community. The chief was clearly more concerned with sanitation and clean water—which he described as vital for public health—than with ICT use, the topic we had come to speak about. After Chief Macha described the challenge of disease transmission in Macha due to poor sanitation and water access, it became clear to us that sanitation, water, and public health are material preconditions of ICT use. It is impossible for those who face serious chronic illnesses or fatal diseases to participate equitably in a global digital society or media culture. These material relations—between sanitation, clean water, and digital networks—are media-related issues we may not have considered if we had not conducted fieldwork in Macha.

In an effort to engage with other people in Macha, we visited the fire camps located outside of the regional hospital operated by the mission (see Figure 7.2). These camps are set on a large expanse of dirt between the hospital gate and Macha's open-air market. The people are there only temporarily, and camp outside in groups so that they can prepare food for and visit their loved ones in the hospital. Those who stay in the fire camps often come from areas beyond Macha's center and throughout the region, and most identify themselves as farmers. Because many of them do not speak English, we worked with a translator. Our translator first approached people on our behalf to find out if we could ask a few questions about their mobile phone and internet use. In some cases, we gave interviewees an instant photo in exchange for their interviews. Some of the participants wanted their photos taken with us. Many of the women we spoke to at the fire camps had never heard of or used the internet (Parks 2015). Some had mobile phones, but did not have enough money to regularly purchase talk time. Thus, they only used the phone to receive calls. Some received calls during our interviews. While people at the fire camps were open to talking to us, these interviews were shorter and less formal due in part to issues of translation and limited digital literacy among our interviewees. However, collectively these fire camp interviews were significant, because they revealed class and gendered dimensions of ICT use or non-use in Macha that were not evident in our interviews with the community's ICT influencers (Figures 7.3 and 7.4).

Ultimately, we wanted to treat our interviewees as collaborators who were contributing dialogically to a kaleidoscope of partial truths and situated knowledges about ICT use in Macha. This dialogic research relationship was not just a theoretical construct that we brought to the fieldwork—many of the people we interviewed turned the tables on us and began asking us a number of questions. This practice suggested that they wanted us to share information about ourselves as well. For example, Chief Macha suddenly began asking us what we studied at our university and what we hoped to gain from our visit to his chiefdom. In turn, a teacher at Macha Central asked us what we intended to do with the information we were requesting and asserted that we needed to do more than simply go home and never return. In both of these cases, our status as white, privileged, foreign U.S. researchers was made the focus of the discussion. These moments left lingering impressions on us and made us question and assess the efficacy of our project in Macha. There was, of course, no way

FIGURE 7.3 Students play outside at Macha Central School during recess.
Source: Photo by Lisa Parks.

FIGURE 7.4   Consider Mudenda and Lindsay Palmer discuss our interview with
Macha Central School teachers.
Source: Photo by Lisa Parks.

our research project could address or resolve the many challenges faced by the
community. Yet our conversations with some interviewees suggested there was
an expectation that it should.

Interviews by Community Researchers

As discussed earlier, at the outset of our fieldwork, we met with our local
partners to introduce ourselves and provide general details about our research
interests. In these talks, we highlighted that we sought to understand how people in Macha think about and use ICTs. After discussing the general focus of our
research project, we taught our partners how to operate Flipcams to record the
interviews. We invited them to conduct the interviews in Chitonga or English,
depending on which language felt most comfortable. We also asked them to interview a broad range of people from different genders, age groups, professions,
and backgrounds. Our partners met each morning and then went in groups of
two or three to sites throughout the community and conducted four to five
video interviews per day. In the late afternoon, the group would reconvene at
the LITA office to transfer and backup video files and work on translations of
interviews conducted in Chitonga. The majority of the interviews conducted
by our Macha collaborators lasted five to twenty-five minutes. Though most
of these partners had been students at LITA, their digital literacy levels varied.
For most, this was the first time they had used digital video cameras to conduct
interviews. Their participation in our research collaboration thus became an
extension of their ongoing digital education (Figures 7.5 through 7.7)

FIGURE 7.5 Project collaborators transfer interview footage from flipcams to computers in a small room we rented near Macha Girls School.
Source: Photo by Lisa Parks.

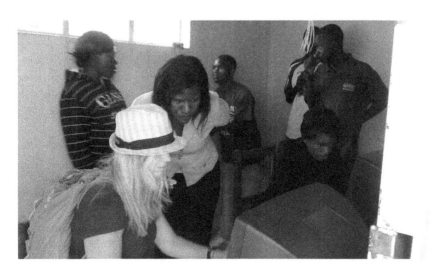

FIGURE 7.6 Research team members worked closely to view and discuss interview footage at the end of each day.
Source: Photo by Lindsay Palmer.

FIGURE 7.7   Project collaborators viewed footage and translated interviews from Chitonga to English.

Source: Photo by Lisa Parks.

Based on our review of these videos, the interviewees sometimes appeared more comfortable speaking with our collaborators in Chitonga. We also learned that our collaborators preferred conducting the interviews in Chitonga, and tended to ask more of their own questions when they used their local indigenous language. This flexibility in language use enabled our partners to move in and out of the language most familiar to them. This was especially helpful, given that they were discussing a technological topic that some perceived as new, strange, or not typically discussed within everyday social interactions in Macha. Moreover, since most of our partners had not collaborated with foreign researchers before, this experience was new for them too. Using their most familiar language made the interview experience more satisfying for the interviewees, allowing greater insights to emerge. This dynamic movement between languages—and between expert/non-expert positions—became a defining characteristic of our collaboration, opening up a transcultural discursive space for thinking about ICTs and fostering more dialogic exchanges.

By adopting a collaborative approach and partnering with Machans, we hoped to accomplish two goals. First, we wanted to learn from a diverse spectrum of people in the community and to explore the variety of thoughts about and experiences with ICTs in Macha. We felt that our community engagement would be stronger if we partnered with people who were already familiar with ICTs and a broad range of social experiences in the community. Second, through the video portion of our research, we hoped to encourage local media education, digital storytelling, and video sharing among Machans. As a result, if broadband Internet becomes more widely available in Macha, community members will not only know how to download audiovisual files, they will also know how to use digital video cameras to produce and share files as content creators. At the end of each of our two fieldwork visits to Macha, we hosted a ceremony in which we presented each partner with a certificate of participation (that could be attached to their résumé) and an honorarium of $100USD. Each partner also received a UCSB T-shirt and a 4G thumb drive as a token of thanks for their contribution to the project. In addition, all of the Flipcams were donated to LITA for their use on future video projects. We also left copies of all of the video interviews in Macha with local project leader, Consider Mudenda.

By participating in our research project, our partners developed or refined their skills in interviewing, video recording, video file transfer, transcription, and translation. They also learned how to work as part of a collaborative research team, provide feedback, and contribute to discussion about the research process. At the end of each of our two visits, partners reported that they not only learned about ICTs in Macha, but also discovered residents' opinions about a host of issues in their community, including the need for better electrification and water, education practices, health care access, and services to address gender-based violence. They also indicated that they were glad to connect with people in the community and hear their thoughts about mobile phones and the internet. In this respect, our collaborative fieldwork served several functions effectively: it provided an understanding of ICT use in Macha; it trained local people to use video, conduct interviews, and discuss technology integration; and it provided an occasion to discuss other issues that were not directly part of the project.

Still, some challenges arose as we attempted to implement our collaborative research design. First, some interviewees wanted to know what would be done with the video recordings and desired something in return for giving an interview. Repeatedly, our collaborators returned from interview sessions and informed us that they had struggled to explain the logic for recording the interviews and to make the interviewees feel as though they were not giving something away without compensation. Some interviewees perceived our partners as perpetrating dynamics of unequal exchange by participating in our project. Thus, our attempt to create a collaboration that would include Machans as researchers and help them to develop certain skills may also have put them in difficult or awkward positions. These experiences led us to contend that there is a need for further research about the cultural meanings of video recording in Macha. Rather than assume that those in rural communities should or want to become content creators, it is important to understand the local histories and dynamics of cultural and political expression. Such work could ensure that visiting scholars do not violate anyone's sense of propriety or safety. We also remind future researchers to think critically about the media they employ to conduct their research as well as the media they are researching, and not to assume absence or uniformity of cultural attitudes toward mediating tools and practices. Since Machans appear to be quite invested in the idea of exchange, and since we did not simply wish to extract information from the community without contributing something in return, we raised funds to bring one hundred solar-powered lights from the non-profit, Unite to Light, to Macha when we returned in 2013. We gave them to interviewees who participated in our research. People were extremely grateful to receive these lights; however, since we could not bring one for everyone in the community, some people ended up disappointed. Consulting more fully with locals before fieldwork about appropriate forms of exchange for participation in research collaboration may help to mitigate miscommunication and disappointment among local community members. However, not every circumstance can be anticipated and not all cultural discrepancies are apparent to every observer. Hence, reflexively pointing to the challenges that researchers encounter during media fieldwork can contribute to a collective knowledge and facilitate stronger future research collaborations.

Second, we also noticed that some of the videos felt a bit contrived when our collaborators were reading our questions to the interviewees, especially when they were speaking in English. Some early videos, which showed our partners reading our suggested questions word for word and acting as if they were outsiders to the local community, even reminded us of Jean Rouch's "ethno-fiction" films such as *Les maîtres fous* (1955). The act of conducting the interviews appeared to be uncomfortable for our partners at times. This was not surprising given that our partners had not developed the questions themselves, were not asking them in their primary language, and were discussing technical topics with which many interviewees had limited familiarity. As mentioned above, we noticed a shift in some of the interviews after our collaborators started conducting the research in Chitonga instead of English. Some of our partners grew more confident with the processes of interviewing and videotaping. This, in turn, put the interviewees at greater ease as well. As more and more discussion occurred in Chitonga, the research collaboration shifted from one initiated by outsiders to one in which local community members were invested.

Site Visits and Path Walking

In addition to engaging with the community and conducting interviews, it was important for us to acquire a sense of the spatial organization of the community. The spatial layout of Macha is a labyrinth of red dirt roads and paths leading in multiple directions. Our fieldwork involved spending time walking along these routes to various schools, the radio station, hospital, open-air market, homes, the bar, the cemetery, and the church. We also visited ICT-related sites— Internet server stations, transmission facilities, Wi-Fi hubs, and power stations in and beyond Macha, in nearby communities Chikanta and Mapanza—in an effort to understand how internet infrastructure was physically organized in the region and how it was functioning. Ultimately, we found that walking led to incidental encounters and observations that were often as significant as the data we gleaned during scheduled site visits or meetings. Walking these pathways also generated insights about the spatial dynamics and mobilities of the community.

For instance, because of the prohibitive costs of gasoline-powered vehicles and fuel, most Machans travel around the community on foot or bicycles. Often, we would see people walking or slowly biking down the red dirt roads with mobile phones in hand. Many of our interviewees noted that mobile phones facilitated a compression of both time and space, because they no longer needed to travel so far to communicate with people located within or outside of Macha. Yet, in Macha, mediated communication still happens alongside and in conjunction with the movement of bodies and information through physical space, down the red dirt roads and paths that, like mobile phones, connect community members to each other. Walkers, bicyclists, car drivers, motorcyclists, minibuses, and truckers all share these roads, and are often engaged in mobile phone use, radio listening, or oral communication as they maneuver through the area. This multiplicity of transit and media reveals some of the particular combinations and materializations of mediated mobilities in Macha. Thus, walking along these paths not only helped us to understand physical distances in the area, but also enabled us to recognize how mobile phone and Internet technologies are integrated with and layered upon already existing Machan transportation and communication practices.

On another occasion, we were walking down the road to reach the Macha Girls School. Just past the school, we found a giant transmission tower that was used by mobile phone providers Airtel and MTN (See Figure 7.8). At the foot of the tower, we met a man who told us that he serves as the security guard and repairman for the facility. He also helped to dig the hole for and install the tower. Sometimes, this man said, he even climbs to the top of the tower on a rickety ladder, in order to maintain the equipment. Our path-walking to this infrastructure site brought forth the human dimension of infrastructural operation—the man who sits at the foot of this tower, every day and every night, to ensure that no trespassing or theft occurs and that the machinery continues to function. This happenstance encounter helped us to better understand how structures of property ownership, labor, and surveillance are interwoven with media infrastructures in rural communities. As such, it is worth thinking about media fieldwork in relation to attunement,

FIGURE 7.8 By walking along paths in the community, we often encountered sites and people of interest, including this massive mobile phone tower and its guard.

Source: Photo by Lisa Parks.

which involves an openness to embodied engagements and unexpected conditions that may compel recalibration or redirection of attention, inquiry, and focus. The concept of attunement recognizes that researchers are embedded within material conditions and affective relations when they are conducting fieldwork. Their phenomenological observations may enhance or complicate empirical findings.

This openness to phenomenological engagement as a source of epistemological discovery is rooted in a rich lineage of feminist thought. We take inspiration from scholars such as Giuliana Bruno, who has approached cinematic viewing as a similarly dynamic, sensorial, and peripatetic experience. Bruno has noted that her book, *Atlas of Emotion* (2002), "mapped out in various cognitive explorations and passing through many different places, is a construction made of multiple passages" (4). She also extols the virtue of haptic knowledge and the value of making reciprocal contact with the materialities of spaces, emphasizing the centrality of the senses in the practices of sense-making. Bruno writes that hapticity is "shown to play a tangible, tactical role in our communicative 'sense' of spatiality and motility, thus shaping the texture of habitable space and, ultimately, mapping our ways of being in touch with the environment" (6). Though she primarily applies this approach to the analysis of films, architecture, and works of art, we found the commitment to mobile attunement valuable for following the circuitous paths of digital networks and media in Macha as well.

Moreover, openness to a broad set of multisensory knowledges that de-center traditional forms of facticity and authority is also connected to the important interventions of feminist critiques of science and technology. Donna Haraway, another theorist who informs our conception of fieldwork, famously deconstructed the illusion of totality behind scientific objectivity, and signaled the impossibility of effacing the positionalities of the researchers. Privileging the concept of "situated knowledges," she has worked to dismantle myths of universality and objectivity. Haraway (1988) writes,

> [O]nly partial perspective promises objective vision. All Western cultural narratives about objectivity are allegories of the ideologies governing the relations of what we call mind and body, distance and responsibility. Feminist objectivity is about limited location and situated knowledge, not about transcendence and splitting of subject and object.
>
> *(583)*

Hence, while our situatedness in Macha can convey some useful insights into local cultural values and digital imaginaries, it also serves to reiterate the inevitable partiality of knowledges that, like the environments in which they are located, are always relative, contingent, and in states of flux.

Conclusion

In the context of this study, media fieldwork involved using methods of community engagement and collaboration, interviews, and site visits to investigate networked communication infrastructures and digital media content generation and sharing in rural Zambia. By collaborating with local partners, we hoped to integrate and prioritize the voices and experiences of Machans in our work with computer scientists who are developing network solutions for rural communities. Though challenges surfaced during the course of our fieldwork, these frictions became core dimensions of our study and altered our perspectives as media researchers. We conclude by briefly summarizing four principles that may be helpful to media scholars conducting fieldwork or building collaborative research projects with developing rural communities.

First, researchers should consider whether and how community members benefit from media fieldwork and research collaborations. Such benefits can range, for instance, from local partners' direct participation in assessing the pros and cons of ICT adoption, to the acquisition of new skills that support local cultural production, to material compensation. It might also include other issues not thought about by the foreign researchers. Integrating community members into the discussion and evaluation of such benefits is vital, even if the project cannot deliver at the scale that is hoped for or expected.

Second, it is important to recognize that audiovisual recording technologies may have multiple meanings and associations in communities where the use of such technologies may be less common. Community members' disposition toward objects such as cameras, audio recorders, and smart phones, and the people who use them, can fluctuate unpredictably between enchantment and rejection, curiosity and mistrust, even if consent to record has been granted. Media fieldwork must recognize the salience of various dispositions toward media technologies, which emerge not only in local media cultures, but also when researchers use media technologies (during interviews or site visits) in an effort to understand such cultures.

Third, media fieldwork can generate unanticipated responses, conflicts, and challenging situations for researchers and community members alike. These effects are part of the uneven structural conditions and transcultural relations that media fieldwork can evoke and produce. While conflicts and tensions should be taken seriously, they are not necessarily insurmountable problems or obstacles. Depending on the issues involved, conflicts can be addressed by the researchers and community collaborators if they are invested in mutual understanding and creating a path forward.

Finally, fieldwork can help media researchers to become more acutely aware of the relationality and contingency of media technologies and cultures in different parts of the world. There is a need to rethink the very terms of media in relation to more multifarious conditions in the world—to investigate the dynamic intra-actions of media within and across various rural communities, materials, practices, and sectors (e.g., water, sanitation, and public health). We hope that our contribution to applied media studies will provoke further discussion of media fieldwork, experimental methods, and transcultural collaborations.

Acknowledgments

This work was funded through National Science Foundation Network Science and Engineering (NetSE) Award CNS-1064821. We are thankful to all of our collaborators and informants in Macha, and especially to Consider and Pharren Mudenda, Elton Muyunga, Austin Sinzala, Acent Milimo, Macdonald Hataana, and Gertjan van Stam. We also thank Abby Hinsman for her contributions to this project.

Notes

1 These UCSB computer scientists are Elizabeth Belding and her former Ph.D. student David Johnson. They had been working with engineer, Gertjan van Stam, who helped to develop LinkNet and Macha Works.
2 In 2012 our local partners were: Ruth Chilweza, Angela Kafute, Nina Kyalifungwa, Trywell Maliko, Peter Miynda, Consider Mudenda. In 2013 our local partners were: Bernard Sishumba, Calvin Muunta, Gracious, Mutinta Maambo,

Peter Miynda, Consider Mudenda, and Evis Muunga. We also consulted in 2013 with doctoral researcher Tony Roberts, of Royal Holloway University of London's Geography Department, who led some video production workshops in Macha in 2011 and 2012.

References

Barney, Darin. (2011) "To Hear the Whistle Blow: Technology and Politics on the Battle River Branch Line," *TOPIA: Canadian Journal of Cultural Studies* 25: 5–28.

Bruno, Giuliana. (2002) *Atlas of Emotion: Journeys in Art, Architecture, and Film*, New York and London: Verso.

Burrell, Jenna. (2012) *Invisible Users: Youth in the Internet Cafés of Ghana*, Cambridge, MA: MIT Press.

Caldwell, John Thornton. (2008) *Production Culture: Industrial Reflexivity and Critical Practice in Film and Television*, Durham, NC and London: Duke University Press.

Couldry, Nick. (2003) *Media Rituals: A Critical Approach*, New York and London: Routledge.

Couldry, Nick and Anna McCarthy, eds. (2004) *MediaSpace: Place, Scale and Culture in a Media Age*, New York and London, Routledge.

Curtin, Michael. (2007) *Playing to the World's Biggest Audience: The Globalization of Chinese Film and TV*, Berkeley and Los Angeles: University of California Press.

Govil, Nitin. (2015) *Orienting Hollywood: A Century of Film Culture Between Los Angeles and India*, New York and London: New York University Press.

Haraway, Donna. (1988) "Situated Knowledges: The Science Question in Feminism and the Privilege of Partial Perspective," *Feminist Studies* 14(3): 575–99.

Huhtamo, Erkki, and Jussi Parikka (2011). *Media Archaeology: Approaches, Applications, and Implications*, Berkeley and Los Angeles: University of California Press.

Imre, Aniko. (2009) *Identity Games: Globalization and the Transformation of Media Cultures in the New Europe*, Cambridge, MA and London: The MIT Press.

Johnson, David, Veljko Pejovic, Elizabeth Belding, and Gerjan van Stam. (2012) "VillageShare: Facilitating Content Generation and Sharing in Rural Networks," *Proceedings of the 2nd ACM Symposium on Computing for Development (Dev 2012)*, Atlanta, Georgia, March 11.

Kraidy, Marwan M., and Joseph Khalil. (2010) *Arab Television Industries*, London: British Film Institute.

Larkin, Brian. (2008) *Signal and Noise: Media, Infrastructure, and Urban Culture in Nigeria*, Durham, NC and London: Duke University Press.

Maîtres fous, Les (1955) Directed by Jean Rouch.

Mayer, Vicki. (2011) *Below the Line: Producers and Production Studies in the New Television Economy*, Durham, NC and London: Duke University Press.

McCarthy, Anna. (2001) *Ambient Television: Visual Culture and Public Space*, Durham, NC and London: Duke University Press.

Michaels, Eric. (1993) *Bad Aboriginal Art: Tradition, Media, and Technological Horizons*, Minneapolis: University of Minnesota Press.

Mudenda, Consider, and Gertjan van Stam. (2012) "ICT Training in Rural Zambia, the Case of LinkNet Information Technology Academy," *Fourth International IEEE EAI Conference on Infrastructure and e-Services for Developing Countries (Africomm 2012)*, Yaounde, Cameroon.

Parks, Lisa. (2016) "Reinventing Television in Rural Zambia: Energy Scarcity, Connected Viewing, and Cross-Platform Experiences in Macha," *Convergence: The International Journal of Research into New Media Technologies* 22(4): 440–60.

———. (2015) "Water, Energy, Access: Materializing the Internet in Rural Zambia," in Lisa Parks and Nicole Starosielski (eds.), *Signal Traffic: Critical Studies of Media Infrastructures*, Champaign: University of Illinois Press, 115–136.

Parks, Lisa, and Nicole Starosielski, eds. (2015) *Signal Traffic: Critical Studies of Media Infrastructures*, Urbana, Chicago, and Springfield: University of Illinois Press.

Pekovic, Veljko, David L. Johnson, Mariya Zheleva, Elizabeth Belding, Lisa Parks, and Gertjan van Stam. (2012) "The Bandwidth Divide: Obstacles to Efficient Broadband Adoption in Rural Sub-Saharan Africa," *International Journal of Communication* 6: 2467–91.

Spitulnik, Debra. (1998) "Mediated Modernities: Encounters with the Electronic in Zambia," *Visual Anthropology Review* 14(2): 63–84.

———. (2002) "Mobile Machines and Fluid Audiences: Rethinking Reception through Zambian Radio Culture," in Faye D. Ginsburg, Lila Abu-Lughod, and Brian Larkin (eds.), *Media Worlds: Anthropology on New Terrain*, Berkeley: University of California Press, 337–54.

Srinivasan, Ramesh. (2017) *Whose Global Village? Rethinking How Technology Shapes Our World*, New York and London: New York University Press.

Starosielski, Nicole. (2015) *The Undersea Network*, Durham, NC and London: Duke University Press.

van Stam, Gertjan. (2014) "Community-Enabled Networks: Macha Works and its Strategy for ICT Access in Rural Areas," *Participatory Networks Workshop at PDC 2014*, Windhoek, Namibia.

———. (2011) *Placemark: Observations from Rural Africa*, Macha, Zambia, https://drive.google.com/file/d/0B_AoSsevFs-uR2huV05LYjVWT1E/edit

Venegas, Cristina. (2010) *Digital Dilemmas: The State, the Individual, and Digital Media in Cuba*, New Brunswick, NJ: Rutgers University Press.

Zheleva, Mariya, Arghyadip Paul, David L. Johnson, and Elizabeth Belding. (2013) "Kwiizya: Local Cellular Network Services in Remote Areas," *MobiSys'13*, Taipei, Taiwan, June 25–28.

8

RAPID RESPONSE

DIY Curricula from FemTechNet to Crowd-Sourced Syllabi

Elizabeth Losh

Acknowledging the legitimacy of the emotional vicissitudes of the *Zeitgeist* online now constitutes so much of the 24/7 work cycle of applied media studies in the field of digital culture that it has become something of a potential occupational hazard for scholars seeking professional advancement in the academy. Although earlier Internet critics could dismiss these fluctuations of powerful affect as temporary "flame wars" or "blogspats" from a position of theoretical remove, feminist interventions by necessity must be empathetic, copresent, and available to perform a variety of forms of affective labor in the wake of collective trauma, outrage, or perturbation. As Hardt and Negri (2000) define it, affective labor can be extremely extractive in character, although it is often uncompensated, feminized, and outsourced in the global market. Now that methodologies of participant observation and situated criticism have become more common in media studies with the rise of science and technology studies (STS) in the field, the application of theory to practice can also come with high emotional costs, as feelings of being buffeted by the forces of intense attraction or repulsion may discomfort researchers more accustomed to static poses of placid neutrality, and new obligations for reciprocity and responsibility to informants may require expending limited resources in the form of affective labor dispensed to a range of communities of practice serving as co-researchers.

This essay considers how a model of rapid response with collaborative online pedagogy is increasingly becoming a normative practice in applied media studies and urges institutions in higher education to validate both the intellectual work and the affective labor that faculty contribute to public humanities efforts. It also calls upon the academy to recognize open syllabi as both a genre of cultural expression and a form of scholarly publication that depends on reaping the benefits of the social capital and digital literacy of junior scholars, which in turn was built by sustaining interpersonal networks and adopting an active stance of constant listening to multiple interlocutors.

My own experience with the open syllabus paradigm began with the formation of FemTechNet, the project for an international multilingual curriculum based on video dialogues about feminism and technology on topics such as digital labor or cyborg identities, and keyword videos that might emerge from these discussions as boundary objects (Losh 2012a). This initiative grew out of preliminary conversations between Anne Balsamo and Alexandra Juhasz in 2011 and soon included hundreds of applied media scholars from a variety of disciplines who were enthusiastic about participating (Losh 2012b). Balsamo and Juhasz also were channeling strong emotional reactions to the rise of Massive Open Online Courses (MOOCs) offered to tens of thousands of students by elite institutions like Stanford or Harvard with learning management systems that tracked progress through video lecture modules and used either automatic assessment—often of multiple-choice questions—or pre-templated peer review by fellow students.

Many feminists were outraged by what they saw as a publicity stunt by MOOC creators that undermined basic principles of critical pedagogy, student empowerment, and consciousness raising, and they were also irritated by the giddy enthusiasm of administrators and advocates for neoliberal efficiencies and entrepreneurialism for these seemingly free courses. In response, this feminist cohort of technology scholars banded together in FemTechNet to share syllabi and resources in what was deemed a DOCC or Distributed Open Collaborative Course. Maintaining the FemTechNet network has required dedicated activities of intense community management and emotional support by its co-facilitators and by members of the multiple committees (including the critical Operations Committee). Mourning, celebrating, squabbling, embracing, and releasing and constraining powerful feelings in our community via online teleconferencing technologies took place in an environment that was mediated, hypermediated, remediated, and premediated in the extreme.

Although the goals of open educational resource initiatives (OER) may sound laudable, big data approaches like the Open Syllabus Project (2014–2017), which has scraped information from over a million syllabi, may raise troubling questions for instructors about informed consent, pedagogical privacy, and quantified metrics with its algorithmically generated "teaching score" for assigned works. In contrast, the FemTechNet (2013–2017) DOCC operated according to an opt-in principle, determined by voluntary association with the feminist collective and the choice to self-publish as little or as much curricular material as an individual contributor might deem appropriate.

"The Selfie Course" in 2014 was my next experience with developing an applied media studies curriculum collectively. In this case, the course depended upon the leadership of Terri Senft, who formed the Selfie Researchers Network (2014) as a Facebook group (now comprising over three thousand members), and Alice Marwick, who set up the website for the twelve core researchers who created the Selfie Course and the Selfie Syllabus as a specific instantiation of the aggregation of research and researchers on digital self-portraiture, mobile devices, and social networks (Losh 2015a). I was paired with Jill Walker Rettberg of the University of Bergen to produce a unit on "Dataveillance, Biometrics and Facial Recognition." We were all encouraged to use scholarly readings from open access publications and to make assignments accessible and flexible, so that individual instructors could adapt them for their own classroom use, and learners outside of the academy could try the exercises out at home and with friends in their own communities.

Like FemTechNet, the Selfie Researchers Network incorporated many techniques from an applied media studies rapid response team, given a similar need to process a barrage of media attention to the scholarly topic. Just as 2012 was declared to be the "Year of the MOOC" by the *New York Times* (Pappano 2012), in 2013 "selfie" was deemed to be the "word of the year" by the *Oxford English Dictionary* (OxfordWords 2013). Hundreds of news stories ran on the topic, and many of them emphasized the supposed vanity and superficiality of young selfie-takers. If FemTechNet acknowledged the presence of negative affect in response to a perceived MOOC love-fest in STEM education, the Selfie Researchers Network attempted to effect the opposite kind of affective reversal by countering what they perceived of as a blame and shame response from mainstream culture against female friendship networks and communities of color. As Senft said in an interview, "[t]he impetus for young girls to make media is strong, as it is for the world to push back with misogyny and slut shaming" (Losh 2015b). She denounced how the "general head-wagging and indifference" of "this larger conversation" often "ends up with the death of young girls."

Senft also noted the centrality of selfie production to the Black Lives Matter movement that has protested killings of unarmed civilians by police (Losh 2015b). The shooting of Michael Brown in Ferguson, Missouri in 2014 also spurred the creation of the #FergusonSyllabus (Chatelain 2014), which used the power of hashtags as aggregating metadata to collect contributions from educators. Georgetown history professor Marcia Chatelain (2014) remembered her own experiences of powerful affect during the civil unrest in Los Angeles in 1992 in response to the acquittal of the officers involved in the beating of Rodney King and feeling "so overwhelmed by all of it." As Chatelain explained in an interview with her campus' *College News*, "I felt it was important to create a way for other educators concerned about how students understood what happened, and I simply want people to commit to thinking about ways to talk about it" (Georgetown College 2014). Core readings were derived from materials that Chatelain was already teaching in her classes about "urban history, white flight, school desegregation, unrest in the 1960s, and the dynamics of race and political participation." Further iterations included children's books to make materials relevant to K-12 educators. Another version of the Ferguson Syllabus emphasized contributions by sociologists (Krutka 2014; Sociologists for Justice 2014). Using a similar donation model as the one that launched #FergusonSyllabus, Chatelain urged participants to contribute print works from the syllabus to under-resourced schools to create a Ferguson Freedom Library.

This transition from online collections of digital ephemera to print artifacts was also important for Chad Williams of Brandeis University, the creator of the #Charlestonsyllabus (2015). Now available as a print anthology from the University of Georgia Press with the title *Charleston Syllabus: Readings on Race, Racism, and Racial Violence* (2016), Williams' book with Kidada Williams and Keisha N. Blaine commemorates the 2015 mass shooting of black parishioners by a white supremacist at a historically significant African Methodist Episcopal Church, which counted Denmark Vesey among its founders. As Williams comments on the website, "#Charletonsyllabus is more than a list. It is a community of people committed to critical thinking, truth telling, and social transformation" (2015).

The #OrlandoSyllabus, which was published online by the queer theory blogging collective *Bully Bloggers* under the byline of Dartmouth professor Eng-Beng Lim (2016), attempted to provide a way for educators to help students process another mass shooting. News accounts of the Orlando massacre that took place at the Pulse nightclub in 2016 unleashed deep fears about heightened homophobia and Islamophobia, which were manifested in shocked reactions from those who viewed gay clubs as social sanctuaries, sites of coming of age, and communities of knowledge production. Instructional units in the syllabus were organized around theoretical topics that incorporated readings from Judith Butler, Hélène Cixous, Audre Lorde, Jacques Derrida, and many others. Readings covered subjects such as "From Gender to Gun Performativity," "Surviving Killabilities," "Laughing at Masculinist Rage, Corruption and Mass Shooting," and "Getting Toxic and Terrifying." In addition to scholarly essays, the syllabus recommended blog posts, opinion columns, graphic novels, plays, poems, and music playlists. It also included specific political criticisms of the anti-immigrant rhetoric of Donald Trump and the intransigence of the gun lobby, and did not shy away from the kinds of messages of advocacy that are generally forbidden in the supposedly neutral pedagogical spaces of public higher education. As part of a larger #PulseOrlandoSyllabus and #OrlandoSyllabus effort, the Digital Library Federation (2016) tallied nearly forty pages of resources compiled in only twenty-four hours.

It could be argued that this tendency for rapid response and crowd-sourced content might eventually undermine the gravity of academic ethos and its traditional gatekeeping and legitimating functions. For example, hashtag syllabi have been created to mark a broad range of events in popular culture, from the debut of an album by a top forty artist in #LemonadeSyllabus (Benbow 2016), to the launch of a mobile augmented reality game for smart phones in #PokemonGoSyllabus (Hustad 2016; Massanari et al. n.d.). Nonetheless, I would argue that this work of translation, adaptation, creative appropriation, and community organization is becoming increasingly significant in the academy, and that the hashtag syllabus epitomizes what the vanguard of applied media studies might aspire to achieve. Furthermore, like many feminists, I value participatory scholarship that undermines conventional divisions between research, teaching, and service in the academy. At the same time, I acknowledge that these tactics can drain time and energy from junior faculty—particularly pre-tenured faculty of color who are most vulnerable to the precarities of the promotion process—and that their affective labor merits greater compensation than they currently receive.

As a case in point I interviewed Roopika Risam (2016) about her experiences creating the #BrexitSyllabus that responded to anxieties related to the 2016 popular ballot initiative that severed decades of political and economic ties between Great Britain and the European Union. Risam explained that her interest in maintaining both critical distance and empathetic identification could be traced to her own position as a dual national: "I am a British citizen and a US Citizen." She described herself as "deeply invested in Brexit not passing" and recounted how she "stayed up all night increasingly upset and increasingly frustrated." At the same time, as a scholar she could see "patterns" worth interpreting that might build on her "work on black Britain and Asian youth movements in England," given how "former industrial sectors that had been decimated" might differ from the more privileged urban milieu of "my sister who works in finance in London."

As Risam was able to make "connections between what I was seeing in the media and the academic side of things," she experienced positive affective states from the gratification of being able to apply useful theoretical lenses to the chaos, grief, anger, and mourning expressed by members of her online social networks. She continued, "I was so glad that I had spent so much time learning about contemporary British politics and cultural material, so I could process what was happening in real time. I had a theoretical framework, knew the historical background, and had read the literature that led up to this horrible political moment. For me, I was interested in how I could make myself feel better" in the environment of "anger, unhappiness, and anxiety that I was feeling" and "that I could share with the public based on my own training." Thus, for her the reparative work began with assuaging her own feelings in her sense-making efforts, which she could then pass on to colleagues and to students who might need to process their own negative emotions.

Risam explained that in designing the #Brexit hashtag syllabus, she was thinking about models like the #Charlestonsyllabus: "I thought that the hashtag syllabus might be a good way to go in exploring the many historical and economic dimensions and to engage people who are coming at Brexit from other disciplinary knowledges." Relieved from the pressure of defending a comprehensive framework of expertise singlehandedly as an individual scholar, she could "invite other people doing other kinds of scholarship that might help frame this experience." When she checked to see if anybody else had started a hashtag syllabus yet, she received a "snarky and sarcastic" comment from someone who eventually joined the conversation in earnest, despite the person's initial anxiety about establishing distance "too soon." Risam expressed her belief about the cohort that responded that "people weren't always reactive, because they were proactive too."

For Risam this entailed "a different understanding of audience" and an ability to seize an "opportunity" for public engagement. In rhetorical terms this occasion could function as *kairos* that enables speech or even exigence that compels it. She also described "leveraging my own social media network on Twitter" and activating a group of participants, "many of whom were British interested in how race and identity were playing out in Brexit returns. I was tweeting out links that were answered by people I knew who were answering with their own sources." Given the complexity and scope of the syllabus, she was appreciative of generous help from Toneisha Taylor and Hannah Clutterbuck-Cook. Clutterbuck-Cook was a historian rather than a literary critic like Risam or a communication scholar like Taylor. "The three of us were posting a lot of links, and people naturally joined the conversation, which was helpful but also unwieldy. How would anybody make sense of them?"

Platforms like the Wordpress-based *Book Riot* for young adult fiction and social bookmarking were helpful in the process to spread the word, as were authoring interfaces like Storify (CrowGirl42 2016) that were designed to organize Tweets into a narrative arc, but the critical development was a "natural move toward Google docs," where "Hannah and I transferred a lot of links." Unfortunately, this compendium "became a gigantic list of sources that nobody could make sense of" until Clutterbuck-Cook applied her archival organizational skills to "categorizing material to create headers." Faced with the limitation of the Google doc format, Risam credited William Patrick Wend with the formation of a Zotero group as the cluster of researchers grappled with the question of "what kind of format could this take?" As a pragmatic approach, Risam decided to "curate a shorter list of citations on a particular theme" and "start looking at potential platforms that allow paths through material," such as Omeka and Scalar, which are designed for digital humanities publication. It continues to be a work in progress in need of periodic "citizen updates," particularly as the steps in the exit from the European Union are implemented.

Of the resulting syllabus, Risam asserts that "there is a clear political angle" about Britain's "past as a colonialist power" and its status as a nation "heavily inflected by the immigrant populations that started coming in larger numbers beginning in World War II." It is also a scholarly publication that requires the "perspective of an editor." For example, in the "huge category of critical theory," sophisticated graduate training in both intellectual history and political philosophy was required to contextualize and parse complex concepts for the general public. This was a significant investment for those asked to "put in time to explain." Moreover, decisions had to be made about how much background reading to offer, because erudite texts like Georgio Agamben's *State of Exception* (2004) do not generally stand alone in courses about biopolitics, human rights, martial law, terrorism, or refugee crises. Additionally, "organization and length are key challenges when inviting anyone who wants to be involved." In the role of editor, the final version must be "relatively short and clearly explain why a particular source is relevant." When you are offering "a syllabus not sources," there are "constraints that limit" without an "additional layer" that allows students to "dig deeper into actual material."

"This is a way of annotating media," Risam insisted. "Specifically, this kind of hashtag syllabus collects media sources and particularly multimodal sources. This is a way of saying 'this is how you can think through what you are seeing in the media and here are the tools to do that.'" Unlike the computational literacy promised to public audiences by participating in hackathons or inhabiting makerspaces, the hashtag syllabus foregrounds relatively low-tech competencies and mines personal experience and community memory for potentially shareable pedagogical touchstones. It makes no pretense of furthering twenty-first-century STEM education, design culture, or innovative branding in asserting the relevance of the university to constituencies in the general public. It embraces the bricolage and temporary autonomous zones of networked publics to scaffold more ambitious curricula and to promote efforts to advance critical pedagogy and new formats for digital publishing.

The informational labor of collating a hashtag syllabus is also deeply entangled with the affective labor expended by all parties involved in all stages of production and consumption of the syllabus. The labor of care and repair is inextricable from the work of assembling, labeling, and ordering sources and materials, which often includes material from visual culture, such as photographs and films. As Risam explained,

> In this case it was affect that was driving the project. Criticisms were popping up on hashtag. 'Why proceed so quickly? Sit back and wait!' Everyone has their own process, their own way of dealing with an issue that is troubling for them. Affect is part of the project. Affect is part of the criticism, as well as a way to channel our expertise. We incorporate our affective responses to the event into a scholarly project that is public facing and open.

Using intersectional feminist ideas from critics like bell hooks and Barbara Christian, Risam argued for "being very explicit about affect not in Lacanian terms but in terms of lived experience." She lamented the fact that "so often those honest expressions of affect are treated as descriptive or untheoretical." As "forms of applied media scholarly output," such syllabi are generated "in large part by affective labor."

As scholars of color using social media and channeling popular sentiment to apply theory to practice, Risam, Chatelain, Williams, and Lim occupy relatively privileged positions in the academy despite what should be acknowledged as their "high risk/high reward" ventures creating hashtag syllabi. Often such work compromises scholars' sustaining social networks, saps their emotional energies, and makes them vulnerable to trolls and harassers both inside and outside of the academy. Administrators, mentors, supervisors, and external reviewers have an obligation to reward faculty undertaking online collective pedagogical experiments like these high-profile hashtag syllabi seen by thousands of participants. As works of scholarship, they establish a high standard for public accountability.

References

Agamben, Giorgio. (2004) Kevin Attell (trans.), *State of Exception*, Chicago, IL: University of Chicago Press.

Benbow, Candice Marie, ed. (2016) "Lemonade Syllabus." *Issuu*. Last updated May 6, 2016. https://issuu.com/candicebenbow/docs/lemonade_syllabus_2016.

"Brexit." (2016) www.zotero.org/groups/brexit.

"#BrexitSyllabus." (2016) https://docs.google.com/document/d/1RDP55BtP6fRZb DkFC9MGeTwAuch9zFylKit4R9BT5T4/edit#.

Chatelain, Marcia. (2014) "How to Teach Kids About What's Happening in Ferguson." *The Atlantic*, August 25. www.theatlantic.com/education/archive/2014/08/how-to-teach-kids-about-whats-happening-in-ferguson/379049/.

CrowGirl42. (2016) "#BrexitSyllabus Storify." *Storify*. https://storify.com/CrowGirl42/brexitsyllabus.

Digital Library Federation. (2016) "#PulseOrlandoSyllabus, #CharlestonSyllabus, and the Public Good." June 14, 2016. www.diglib.org/archives/12064/.

FemTechNet. (2013–2017) "The Network." Accessed September 26, 2016. http://femtechnet.org/about/the-network/.

Georgetown College. (2014) "The #Ferguson Syllabus." Last updated August 27, 2014. https://college.georgetown.edu/collegenews/the-ferguson-syllabus.html.

Hardt, Michael, and Antonio Negri. (2000) *Empire*, Cambridge, MA: Harvard University Press.

Hustad, Karis. (2016) "This UIC Professor Created a Crowdsourced Pokemon Go Syllabus." *Chicago Inno*, September 15. http://chicagoinno.streetwise.co/2016/09/15/pokemon-go-tips-lessons-studied-in-crowdsourced-syllabus/.

Krutka, Dan, ed. (2014) "Teaching #FergusonResources." *Google Docs*. Last updated July 15, 2015. https://docs.google.com/document/d/1kwZl23Q9tgZ23dxSJWS-WpjZhOZ_mzVPtWL8-pWuLt8/edit?pli=1&usp=embed_facebook.

Lim, Eng-Beng. (2016) "#Orlando Syllabus." *Bully Bloggers*. https://bullybloggers.wordpress.com/2016/06/24/the-orlando-syllabus/.

Losh, Elizabeth. (2012a) "Bodies in Classrooms: Feminist Dialogues on Technology, Part I." *DML Central*, September 2. http://dmlcentral.net/blog/liz-losh/bodies-classrooms-feminist-dialogues-technology-part-i.

———. (2012b) "Learning from Failure: Feminist Dialogues on Technology, Part II." *DML Central,* August 9. http://dmlcentral.net/blog/liz-losh/learning-failure-feminist-dialogues-technology-part-ii.

———. (2015a) "Selfie Pedagogy II: Internet Identity and Selfie Practices." *DML Central,* September 10. http://dmlcentral.net/selfie-pedagogy-ii-internet-identity-and-selfie-practices/.

———. (2015b) "Selfie Pedagogy III: Networked Spaces, Slut Shaming and Putting Selfies in Dialogue with Theory." *DML Central,* October 1. http://dmlcentral.net/selfie-pedagogy-iii-networked-spaces-slut-shaming-and-putting-selfies-in-dialogue-with-theory/.

Massanari, Adrienne, et al., eds. (n.d.) "Pokemon Go Syllabus." https://docs.google.com/document/d/1xYuozfkON-RVZQkr7d1qLPJrCRqN8TkzeDySM-3pzeA/edit#heading=h.2ci4mehk80kn.

Open Syllabus Project, The. (2016) "The Open Syllabus Project—Opening the Curricular Black Box." http://opensyllabusproject.org/.

OxfordWords. (2013) "The Oxford Dictionaries Word of the Year 2013 Is 'Selfie.'" *OxfordWords Blog.* http://blog.oxforddictionaries.com/2013/11/word-of-the-year-2013-winner/.

Pappano, Laura. (2012) "Massive Open Online Courses Are Multiplying at a Rapid Pace." *The New York Times.* www.nytimes.com/2012/11/04/education/edlife/massive-open-online-courses-are-multiplying-at-a-rapid-pace.html.

Risam, Roopika. (2016) Interview by Author, August 1.

Selfie Researchers Network, The. (2014) "The Selfie Course." www.selfieresearchers.com/the-selfie-course/.

Sociologists for Justice. (2014) "Ferguson Syllabus." https://sociologistsforjustice.org/ferguson-syllabus/.

Williams, Chad, Kidada Williams, Keisha Blain, Melissa Morone, Ryan Randall, and Cecily Walker, eds. (2015) "#Charlestonsyllabus." African American Intellectual History Society. www.aaihs.org/resources/charlestonsyllabus/.

Williams, Chad, Kidada Williams, and Keisha Blain, eds. (2016) *Charleston Syllabus: Readings on Race, Racism, and Racial Violence.* Athens, GA: University of Georgia Press.

PART IV
Translation

9

TRANSDISCIPLINARY COLLABORATION AND TRANSLATIONAL MEDIA-MAKING

Kirsten Ostherr, Eric Hoyt, Tara McPherson, Bo Reimer, Lisa Parks, Jason Farman, Elizabeth Losh, Patrick Vonderau, and Heidi Rae Cooley

Kirsten Ostherr

In this chapter, contributors address the core challenges of collaborating across academic divisions such as humanities and science, bridging academic and community practices, and translating between the diverse stakeholders involved in these projects. Much of the recent research on problem-based collaboration and innovation has noted the value of bringing together teams of people with radically different forms of expertise to solve complex challenges (Johnson 2010; Davidson 2011; Ness 2012). Contributors address a common question asked by academics interested in applied media studies, who don't know where to begin: how do you find good collaborators in different, perhaps unfamiliar disciplines? Further, once those collaborators are found, how do you overcome the typical silos of universities structured by departments within divisions like "humanities" and "sciences" to build effective working relationships with your collaborators? How do you translate between fields with radically different training, terminology, and theories of knowledge? What does it take to develop a shared vocabulary?

The motivation for engaging in applied media studies often includes the desire to reach a broader public, but here, too, translation is key. What does it take to translate scholarly practice into public dialogue? How, in practice, does this work? What are some effective techniques for reaching beyond the so-called ivory tower to make meaningful, socially relevant work that matters to academics and to the community at large? What happens when efforts at translation fail? Contributors note that in many circumstances both within and outside of academia, inequalities of various kinds create pressures that need attentive mitigation. Here, the issue of reciprocity is critically important at the interpersonal and the structural or institutional level. For translation to produce meaningful, sustainable results, it must flow in more than one direction. If the need for translation is a side-effect of academic specialization and situated knowledge, then the efforts to collaborate through difference must pay special attention to the interface between participants. Many of the responses to this chapter's queries cite multimedia design as a particularly valuable technique for cultivating those interfaces.

How Do You Find Good Collaborators in Different, Perhaps Unfamiliar Disciplines? Once Those Collaborators Are Found, How Do You Overcome the Typical Siloes of Universities Structured by Departments within Divisions Like "Humanities" and "Sciences" to Build Effective Working Relationships with Your Collaborators?

Eric Hoyt

Media scholars generally have a project in mind when they start seeking out collaborators. If you are based at a university, it's pretty easy to surf your institution's website and identify people in the library and other academic departments who could contribute expertise and add value to your project. The harder and more important step is to think about what your project is going to offer them in return. In most cases, your potential collaborator is already busy with other projects, and his or her salary will remain the exact same regardless of whether he or she agrees to help you. So when you are trying to assemble a team, you should craft a slightly different pitch for everyone you talk with. How can you make your project sound like the most exciting and compelling thing imaginable to that particular person? How can you make them an offer they can't refuse (though without implementing tactics from *The Godfather*!)?

Tara McPherson

Our collaborators come from both in and out the university. Steve Anderson and I began planning for the *Vectors* journal in 2002 and 2003. Our first additions to our small core team were our original co-creative directors, Raegan Kelly and Erik Loyer. We spoke widely to friends and colleagues within the Los Angeles arts community and advertised by word of mouth and through university networks. We already knew of Erik's work as a new media artist and were delighted to be able to bring him aboard. Raegan's talents nicely complemented Erik's, and we were off and running.

As work on the journal began in earnest, we knew we needed backend programming help. This was our greatest challenge early on. We worked with a few different programmers before we found Craig Dietrich, who later became our Lead Technical Designer. From early on, we frequently worked with artists, programmers, and scholars from USC's graduate programs. We have had especially good luck working with programmers and technologists who also have some training in the arts. I'd say the common grounding for our team over the years is that everyone values the arts. This gives us a shared base even as different team members have very different skill sets and roles within the team.

Bo Reimer

My situation has been very fortunate, since I was part of creating the multidisciplinary School of Arts and Communication, which was founded on the idea of creating a school to which teachers and researchers came precisely because they wanted to work multi- or trans-disciplinary. So finding good collaborators have been simple within my own school. In addition, my university—Malmö University—is fairly small, which also makes it easier to find possible collaborators. I believe it also helps that this is a young university, where the interest in collaborating and experimenting may be greater than in older universities.

Lisa Parks

I have found strong collaborators in the past by participating in interdisciplinary lecture series on my campus. A talk I gave on mobile telephony in Mongolia led to a meeting with a computer scientist several years ago, and we have been writing federal grants together and collaborating on research projects ever since. We also expanded our team and added more partners and disciplines to our projects. There are far more grant opportunities for computer scientists than film and media studies scholars, yet we have managed to find ways to conceptualize projects that bring our fields together. Collaborations are never perfect or easy. I have learned that it is important to enter them with modest expectations rather than high hopes. Collaborations across disciplines also take much more time and energy than working alone. These relations can be sustained by learning a few things about each other's disciplines, regular face to face meetings, exchange of research findings and design solutions, development of presentations together, co-authoring of articles, and so on. Sometimes I wish campuses would organize speed date-like sessions between researchers in the humanities and sciences as I think there are more shared research interests and potential for collaboration than we are aware of.

Jason Farman

Interdisciplinarity has taken many different forms in the academy. At the University of Maryland, I've been lucky to be involved with various centers on campus that function independently of departments and the constraints of disciplinary silos. For example, the program I direct, called "Design Cultures &

Creativity," draws on students and faculty from across the University, including the sciences, the arts, and humanities. Similarly, I work closely with the Human-Computer Interaction Lab (HCIL) and the Maryland Institute for Technology in the Humanities (MITH). These two centers bring people together from different units on campus to collaborate and engage in interdisciplinary research. Having worked at a broad range of institutions in my career, I can attest that seamless collaboration across departments and with faculty in other units is not always a structural priority for every university. At the University of Maryland, while it is easy to work across disciplinary boundaries, faculty in different fields have different teaching requirements; this leads to wide variability in the kinds of time that faculty have to devote to scholarly publications or to undergraduate teaching. So, structural inequities across different fields at a university often inhibit the amount of time that collaborators can devote to the kinds of projects we may want to pursue. As large teaching loads or administrative duties are placed upon some faculty, the amount of time, energy, and resources that a faculty member can give will vary from unit to unit. This, to me, is one of the largest challenges of doing applied media studies with researchers in different disciplines.

Elizabeth Losh

Because the collaborations of FemTechNet, Selfie Researchers, and Connected Courses all focus on pedagogy, we are often interacting with pre-disciplinary students who might be relatively open to scholarship that cuts across multiple fields. In the case of my situation at my current institution, which is a liberal arts university, I have benefitted from a new curriculum with general education courses that mandate that one discipline "reach out" to another. This makes it easier for me to teach collaboratively with materials from the humanities, the arts, the social sciences (for ethnographic literature about human-computer interaction), and computer science in a gender and technology course.

Patrick Vonderau

A good way to find collaborators in other disciplines is to invite them to transdisciplinary workshops or into research project collaborations. This normally starts by developing a research bibliography on your subject that includes work in other fields. The hard part is to identify potential key contributions in disciplines you are not trained in yourself. I normally begin by contacting colleagues from other faculties at my university and asking them for advice, based on the preliminary research I have already done, indicating where I feel I need expert advice from another community. I don't think there is a major challenge here. In fact, platforms such as Academia.edu, or Researchgate.com (although problematic in other ways), or even Google Scholar create "first windows" of opportunity to quickly identify interesting partners for dialogue. Open Access publishing and scholarly journals, handbooks, and book reviews also greatly help to navigate foreign territory. But these challenges are not particularly new or uncommon, nor unsurmountable. Personally, I like to meet and talk to a lot of people.

Heidi Rae Cooley

Finding good collaborators in different or unfamiliar disciplines involves embracing a logic of adventure and actively seeking opportunities for intellectual conversation and engagement beyond one's usual disciplinary circuits. For me, the Center for Digital Humanities provided such an opportunity, when they invited me and computer scientist Duncan Buell, along with another Media Arts colleague, to apply for an NEH grant that would fund a three-week Humanities Gaming Institute in June 2010. We were awarded the grant. At the conclusion of the Institute, Duncan and I realized that we had a strong rapport, and we were both interested in continuing to work together. But it wasn't until mid-2011–2012, during development of an early modern British social history game, "Desperate Fishwives" (concept by Ruth McClelland-Nugent), that we confirmed that we had a really solid working relationship. We trusted and respected each other; we could be candid with each other (sometimes painfully so); we shared common beliefs about project management; we felt a deep commitment to mentoring students engaged in project development; and we genuinely enjoyed working together. Being "on the same page" and deriving pleasure out of arriving at the same page—the hours planning, trouble shooting, problem solving, and just simply talking—is crucial to strong collaboration. Collaborative work takes time, a lot of time. And while our academic units really do not have a way to acknowledge the amount of time we invest to ensure successful collaboration, we have managed to teach our Critical Interactives course four times now, which suggests that they support the work we are doing.

How Do You Translate between Fields with Radically Different Training, Terminology, and Theories of Knowledge? What Does It Take to Develop a Shared Vocabulary? What Does It Take to Translate Scholarly Practice into Public Dialogue?

Eric Hoyt

I think staying goal-oriented helps with collaborations across very different disciplines. And, by their very nature, applied media studies projects are generally goal-oriented: we know we want to build an app, database, etc. So my advice would be to stay focused on the goals and outcomes but try to understand the processes that your collaborators will need to follow to get there. If you are trying to create an app, for example, you don't need to be masterful at coding in Java or Python, but you should have a basic understanding of how algorithms and computer programming work.

Who are the audiences for this project? And how are we adding value to their lives? These are crucial questions that need to be addressed early on in any applied media studies project. If you can articulate precise and clear answers, then translation and outreach become much easier—you find the online communities in which your audiences spend time and you show them you have something that they will find useful and interesting. This can start a dialogue and process of engagement that continues for years, as I have experienced with the Media History Digital Library and Lantern. But it all starts with knowing your audiences and trying to add value to their lives.

Tara McPherson

In building a team that could work across disciplinary lines, we found it important to foster a shared vocabulary. Words like "theory," "archive," or "ontology" can mean different things across different domains, so we try to create an environment where everyone feels free to ask for clarifications. As we began to work in intensive and iterative collaborations with scholars from a variety of disciplines, it was also helpful to have clear ground rules. Every project started with a lot of conversation and shared background materials to help get everyone on common ground. Scholars would bring in materials that were relevant to the project they had in mind, and we'd spend a good deal of time with white boards mapping out possibilities. We also developed a set of guiding questions that scholars would answer in advance of our first meeting, and this framework offered a useful launching point for the collaborations that would follow.

Scholars also needed to know that our design, editorial, and programming team was not working *for* them but *with* them. At the start of each collaboration, an editor (Steve or I) served as a 'translator' between the domains of the scholar and the artist. We were more heavily involved in the early meetings until a common vision and language began to emerge. Our lab also sketched out a development process that sometimes suspended the relentless interrogation that can be part of humanities scholarship in order to allow room for experimentation and making. Theory and practice often operate through very different modalities. Our end goal was to bring them into productive tension, but you sometimes need a space where one mode or the other is set aside momentarily.

Not all the work of our lab is meant to reach a broader public, but a good deal of it does. A good example of a project that circulates in different realms is *Public Secrets*, an audio documentary created by Sharon Daniel and Erik Loyer as part of the *Vectors* fellowship process. The piece uses audio testimony from incarcerated women to make an argument toward the abolition of prisons. It's a very experimental piece, driven by an innovative interface, sound design, and thematic elements. It brings together the women's voices with theoretical work on human rights, the carceral state, and feminism. The piece has circulated to a wide variety of audiences, including academics, international museums and festivals, and activist organizations. It was even an Honoree for the Webby Awards. I think it was able to move between these audiences because it moves beyond text and invites the listener to engage deeply with the testimony she is hearing. So, one way to reach beyond the academy is to take issues of multimedia design quite seriously.

In launching *Vectors*, we were exploring how you might immerse yourself in an argument like you get immersed in a film. We sought to invite interactivity via interface design. We wanted to see if popular digital forms could be meaningful for scholarship. We were pleased that these experiments also sometimes reached a broader audience. In developing the authoring platform Scalar, we intentionally created a free, open-source tool that anyone could use. While it was designed primarily for academic use, it has been taken up outside of the academy as well.

Bo Reimer

Developing a shared vocabulary takes time and commitment. "My" School of Arts and Communication has now been in existence for almost twenty years, and there is a sustained, ongoing attempt to develop, and keep alive, a common language between teachers and researchers with backgrounds in media, arts, design, and technology. Attempts to do this include having research retreats focused on discussing what is meant by collaboration, by combining theory and practice, and so on. It includes attempts to write joint research applications. It includes co-teaching in courses combining theory and practice. Both time and commitment are important, but so is interest! If the interest is not there, then time and commitment do not help.

A traditional way of reaching a broader public is of course to take part in public debates through the writing of newspaper articles. Having a webpage which aims at reaching non-academics is another way. That is how we try to work with the webpage for the research lab Medea that I direct (http://medea.mah.se/). We also produce a podcast—Medea Vox—with topics of more general interest (http://medea.mah.se/vox/). And we arrange a seminar series—Medea Talks—which has both academic speakers and speakers from the media industry, and which draws not only academics but also the general public (http://medea.mah.se/medea-talks/).

However, it is important to see connections to people outside of academia not only as the arranging of events. It is as important to conduct research involving non-academics. For our sake, that is done through the creation of so-called Living Labs, where we work with civil servants, NGOs, companies, and citizens (http://medea.mah.se/malmo-living-labs/).

Lisa Parks

I approach the work of translation primarily by listening. I may explore paradigms, terms, and theories from different fields, but I do not aspire to master them. It is challenging enough to try and immerse oneself in a single discipline in a way that recognizes its breadth and depth, much less attempt this with multiple fields simultaneously. I tend to not work to create a shared vocabulary—that process can seem somewhat contrived to me. Instead, I try to value the differences that collaborators bring to the table as well as the miscommunication and knowledge gaps that come from working across fields. When building collaborative relationships, it is often helpful to read or watch something together and discuss it, or talk about how those in different disciplines might approach a particular research question or solve a problem.

How to translate scholarly practice into public dialogue is a vital question. It may be helpful to think about the multiple forms that scholarly research findings can take. Although many of us work in institutions that value the book and the peer-reviewed article, universities also value public service and engagement. It is possible to share research findings in op-ed pieces, in museums and art galleries, on websites, and in public meetings. I once gave a talk at the YWCA and had one of the most inspiring Q&A sessions of my career. My colleagues have testified as expert witnesses in state and Supreme Court hearings and at the FCC. An applied media studies perspective involves conceptualizing the university as open and integrated with different publics in the U.S. and abroad.

Jason Farman

I am a humanities scholar who often collaborates with people in the field of human computer interaction (HCI). What I've found is that many of the terms we use have different connotations and resonances in different fields. Similarly, I find that much of the work done in media studies is simultaneously being done in other fields like HCI, though these fields may not talk to each other very often. Instead, the fields tend to duplicate research or simply speak in parallel ways that rarely intersect. Finding ways to put these fields in conversation is one of the largest tasks of interdisciplinary research. As I have pursued this kind of work, it has required that I take on new methods from different fields. This has given me access to modes of exploration and research that I had not been exposed to until recently. Thus, working collaboratively with people in different disciplines has expanded my own abilities to answer the questions I am most curious about.

I believe that all faculty members' work should translate to a broad audience and reach that audience through venues that are appropriate for the work. I think "translation" is a good term for this kind of work. Complex and abstract thinking play important roles in the kinds of research that we do; that said, most of the work we do can be translated and iterated for a broad audience. Thus, I think we can do both. I believe that we can write, publish, and produce work that addresses our disciplinary colleagues using field-specific language, work that is complicated, abstract, and nuanced. But we can also take those findings and ideas and present them in ways that are accessible to a very broad audience and can ultimately impact how that audience thinks.

There are a couple of challenges for this kind of work. First, academics are rarely trained to write and produce work in genres that are appropriate for broad audiences. Second, the tenure and promotion process rarely gives the appropriate value to this kind of public scholarship in ways that it should. One assignment that I've been doing lately with my graduate students is to have them identify a piece of public writing that fits the kind of voice they can imagine using in their own work. Then they take a case study or object that they are working with and begin to mirror the prose of their chosen author. By mimicking the style, structure, and approach that these authors take, my students begin to uncover how these authors accomplished their sophisticated public writing. This mode of production could be replicated for a wide range of scholarly outputs, including hands-on work. I do something similar when I have my students design their apps. By identifying what works or what does not work in existing apps, they begin to see how the choices others have made can teach them about the choices that they, too, have to make.

Elizabeth Losh

Translating across fields is often difficult. The "applied" part of applied media studies—when a group of academics is attempting to serve as useful resources to a classroom, community space, policy meeting, or institutional archive— requires that participants use a shared lexicon of accessible vernacular terms and avoid the jargon of their specializations. But for the service to count as research it has to be translated back into specialized scholarly discourse, compounding all the regular challenges of collaborative writing. For example, some scholarly collaborations assume common genres for the publication of results that include a formal "methods" section, which is an unfamiliar custom to me, because this is rare in humanities scholarship, which focuses on interpreting texts.

As I write these answers I find myself reflecting on a terrible week in which public anger about police shootings of black men can't be adequately addressed by academic talking heads. It is in these times that applied media studies seems insufficient, particularly when digital activism doesn't seem to be changing the statistics. I already often feel like an outsider to the practices of hashtag activism, witness journalism, live streaming, and self-narrowcasting appearing in my feed, and I am reticent to pontificate or claim competence as a translator. The news this morning about the killing of five police officers by an enraged sniper on the same Dallas street where I walked between academic conference events a few months ago is a particularly jarring reminder of the life of privilege I lead.

Patrick Vonderau

This depends on the kind of translation that really is required. Do we always need a shared vocabulary? This ultimately depends on the site, format, and expected result of our exchange. We are under no pressure to develop these kinds of projects and largely free in defining where we want to go. Also, our exchange is not necessarily transactional—as implied by Peter Galison who famously borrowed the notion of the "trading zone" from anthropologist Nicholas Thompson to describe how, at the border between various academic subcultures, new knowledge evolves in the coordination of various values and meanings. Transactional exchange always is directed towards concrete results that are economical or can be economized: re-used, patented, turned into projects, etc. "Applied media" work, on the other hand, allows us to experiment with terminology and foreign knowledge. For instance, we may take up a descriptive or metaphorical term used in another community of practice and turn it into an analytical term. Just think of the word "algorithm." The challenge I see is that such foreign or uncommon vocabulary sometimes may turn into jargon or into a signifier of something else, namely, our alleged proficiency in speaking foreign languages.

Translating scholarly knowledge into public dialogue obviously is not a challenge specific to applied media work. And vice versa, doing applied media work is not necessarily the best or preferred solution to the problem of communicating subtle distinctions and detail to an audience that seeks news and entertainment. At Swedish universities, such communication to the broader public is called the "third task" (*tredje uppgiften*) alongside research and teaching, and required for anyone seeking promotion. I have given interviews to newspapers, radio, and television, among others, but always found the expert role to be a challenging one. In my view, *The Conversation* (theconversation. com) is a format better suited to demonstrate the social or political relevancy of humanities work.

Heidi Rae Cooley

One of the most productive things Duncan and I did early on in our collaboration was to sit in on a course in each other's disciplines. In spring 2011, I sat in on "Algorithmic Design II" (CSCE 146), which Duncan was teaching. The following spring, Duncan sat in on my Film and Media Theory (FILM 473) course. Whereas I learned how to think about data structures (e.g., arrays and linked lists, stacks, graphs, trees) and was introduced to the Traveling Salesman Problem (a discovery that proved significant for my book *Finding Augusta* and its digital supplement *Augusta App*), Duncan made his way through various theoretical texts about, for example, mediation and how media objects and technologies position and train audiences. These experiences served to better prepare us for "hearing" how students enrolled in our team-taught Critical Interactives class think. Additionally, we spend many hours discussing the direction a project is taking and "next steps"; this we frequently do over food and beverage. Finally, we have made a concerted effort to co-present at conferences (including Society for Cinema and Media Studies and HASTAC) and other public venues, and we have co-authored several articles (in addition to the writing we do in pursuit of grant funding).

Translating scholarly practice into public dialogue requires that one be in ongoing conversation with members of the public or community. This means being present and visible in the community, attending meetings and events, and participating in activities. It means listening, really listening, and being responsive to individual and collective interests and concerns. It also requires time and patience. One has to develop trust—and demonstrate (repeatedly) that it is deserved.

In the case of the Ward One project, Duncan and I have spent approximately seven years cultivating what, I think, has become a deep and respectful relationship with colleagues, students, and former residents of Ward One. We first approached historian Bobby Donaldson in 2008 about any possible interest—his and the Ward One Organization—in the development of a geo-locative mobile application featuring the history of the old ward. Our first meeting with representatives of the organization happened in June 2014—after we had team-taught two courses (fall 2012, spring 2014) that produced a prototype application for iPad, *Ghosts of the Horseshoe*, that presents the unacknowledged history of slavery that made possible the historic Horseshoe, heart of the University of South Carolina's Columbia campus. The *Ghosts* project demonstrated (1) how we work with colleagues and students, both those whom we teach and those whose robust scholarly website we were translating into a geo-locative application, and (2) how we approach topics of a politically sensitive nature. At the same time, Duncan and I have developed relationships with representatives of South Carolina Educational TV (SC-ETV), Historic Columbia Foundation, and the Nickelodeon Theater, who attend Critical Interactives end-of-term presentations-demonstrations, raise questions, and offer insights. These efforts to engage with community organizations have expanded our "reach" and made for richer conversations about how the Ward One project might evolve and what it might accomplish.

However, our real work, in this regard, is the "smaller," less notable work of being in conversation with the Ward One Organization. Mattie Roberson, one of the central voices in the Ward One app (to date) and president of the organization, emails us regularly—in one instance, to request that we convert the 2015 end-of-semester program for the public presentation of the Ward One app into a brochure for Ward One Organization members to distribute at the February 2016 Black History parade (which we did). We've attended Ward One Organization meetings, as well as the most recent Ward One Organization Biennial Reunion (fall 2015). And we're currently in conversation with Richland Library Adult Programs staff about developing a memoire writing program and/or an oral history program, so that former residents can "capture" their stories even when the Critical Interactives class is not being taught. Here, their interest in recording their histories is entangled with our interests in including their voices in the app, of course; but also, it's very much about continuing the conversation.

References

Davidson, Cathy. (2011) *Now You See It: How Technology and Brain Science Will Transform the Way We Live, Work and Learn*, NY: Viking/Penguin.

Johnson, Steven. (2010) *Where Good Ideas Come From: The Natural History of Innovation*, NY: Riverhead/Penguin.

Ness, Roberta. (2012) *Innovation Generation: How to Produce Creative and Useful Scientific Ideas*, NY: Oxford.

10

COLLABORATING ACROSS DIFFERENCES

The AIDS Quilt Touch Project

Anne Balsamo

How does one name a project, when the boundaries of the effort are indistinct, the time of creation stretches in many directions, and the intentions of designers shift over and over? A provocative name lifts ideas; a narrow one can cloud creative thinking. We happened onto the name for our project through a sideways path while we were discussing the merits of different modes of interaction for the interface design. The name certainly didn't come first, but when it arrived, many decisions were settled. This case study focuses on the creation of a project called *AIDS Quilt Touch (AQT)* (2012), a collection of interactive applications designed to augment the experiences of viewing the AIDS Memorial Quilt (Figure 10.1).

Because the Quilt is a richly textured material artifact, our designs rely on the use of tactile modes of interactivity. Applications have been optimized for display on touch-enabled devices (interactive tabletops, large touch screens, and mobile, hand-held devices) to provide an intimate experience of viewing Quilt information. This modality of interaction is also powerful for viewers who encounter the AIDS Quilt in its textile form. They touch, they are touched, they witness through touch. More than a metaphor, touch is a way of knowing and encountering the other (across time, across life, across networks).

We began this project with a techno-cultural question: how might digital technologies augment the cultural significance of the AIDS Memorial Quilt? In 2013, Richard Kurin, a director at the Smithsonian Institutions, identified an AIDS Memorial Quilt panel as one of most significant 101 objects that defined America in the twentieth century (Kurin 2013). Yet in 2006, newspaper headlines told a different story. On the occasion of the 25th anniversary of the AIDS epidemic, the *Los Angeles Times* sadly announced: "The Quilt Fades into Obscurity" (Zarembo 2006). Even though sections of the Quilt continue to circulate on a regular basis, understanding about the historical significance of this

FIGURE 10.1 Image of AIDS Memorial Quilt on the Mall of Washington, DC, 1996.

Source: Photo in public domain available from http://aidshistory.nih.gov/tip_of_the_iceberg/quilt.html.

extraordinary, ongoing, collaboratively produced cultural artifact is limited. The Quilt belongs to several intertwined histories, including the history of arts activism in the U.S., the history of struggles for gay and lesbian rights, and the history of public health protests. Given that some of the first generation of panel makers are now entering their 80s and 90s, there is rising concern about the future of the development of the archive of Quilt stories. Our project is designed to address the archiving and dissemination of information about the Quilt.

The AQT project has resulted in the creation of a dynamic media system that includes distinct elements such as multimodal content, data structures, interface conventions, and cultural practices. The project unfolded over many years and involved diverse partnerships and collaborations, some of which have persisted, others were more delimited. In this case study, I discuss the AQT project as an example of an applied media studies project.

General Background

The AIDS Memorial Quilt is a unique work of international arts activism that reflects the worldwide scope and personal impact of the AIDS pandemic. From

FIGURE 10.2 Image of Marvin Feldman panel.
Source: Photo in public domain available from http://www.aidsquilt.org/view-the-quilt/
search-the-quilt.

its beginning in 1985, the Quilt has served as an activist tool to provide visual
testimony of the scale of the impact of the HIV-AIDS pandemic. As a collection
of individual panels, it is the largest textile quilt in the world. It simultaneously
functions as an intimate tribute to lives lost, a form of political activism, and a
living memorial that continues to educate, commemorate, and celebrate what
it means to live in the age of AIDS.

When the Quilt was first displayed as part of the 1987 National March on
Washington for Lesbian and Gay Rights, it included 1,920 panels. Since then,
the Quilt has been displayed on the Washington Mall several times: in 1988,
1992, and 1996. Under the stewardship of the non-profit organization The
NAMES Project Foundation (NPF), Atlanta, the Quilt now comprises 48,000
individual panels that commemorate more than 98,000 names (Figure 10.2).

This represents roughly 15% of the number of people who have died of HIV/
AIDS in the U.S. The size of the Quilt is staggering. Each panel of the Quilt
measures three feet by six feet; every panel is stitched into a 12-foot by 12-foot
block (Figure 10.3).

If the Quilt were laid out for display it would cover more than 1.3 million
square feet. If a person spent only 1 minute visiting each panel, it would take
33 days to view the Quilt in its entirety. The impact of the Quilt plays out
at different scales; certainly its cultural significance is tied to its massive size,
the quantity of names represented, and the spatial dimensions of its array. But
its impact also plays out at the scale of individual panels, where the stories of
tens of thousands of people—those who died and those who lovingly created
the panels—are literally stitched into a historical material archive. Imagining

FIGURE 10.3   Image of eight panels that became the first block of the AIDS Quilt.
Source: Photo in public domain available from http://www.aidsquilt.org/view-the-quilt/
search-the-quilt.

an innovative way to tell these stories was the initial motivation for the AQT
project.

In 2001, my colleagues at Onomy Labs—Scott Minneman, Mark Chow,
and Dale MacDonald—approached the NAMES Project Foundation to collaborate on the creation of a browsable image of a virtual Quilt. Since the early
1990s, the NPF has maintained a website that includes a simple search function
that allows a user to find a particular panel using the block number or the name
on the panel (Figure 10.4).

NAMES staff members had long realized that this limited search capacity
did not provide a robust experience of the AIDS Memorial Quilt, but it was the
best that the foundation could do with limited resources. Although we didn't
realize it in the beginning, one of the critical phases of the AQT development
effort was to create a robust backend database to manage Quilt content, images,
and meta-data. Backend development takes resources, so while we understood
what the project would require, we were unable to actually begin the effort
until we found funding for the database creation.

FIGURE 10.4 Image of NAMES Project Foundation Quilt web search page.
Source: Photo in public domain available from http://www.aidsquilt.org/view-the-quilt/search-the-quilt.

Over the next decade, my Onomy colleagues and I worked to find funding to create digital applications to enable people to experience the AIDS Quilt. We approached several corporate and national foundations, but the project was always just a bit "off" topic. It wasn't until the National Endowment for the Humanities created the Office of the Digital Humanities in 2006 that we could seriously consider pushing the project forward. It had taken nine years for culture to catch up with our ideas for AIDS Quilt Touch. In 2010, we received an NEH Digital Humanities Start-Up Grant that enabled us to create the first version of a browsable Virtual Quilt.

Our initial (and abiding) objectives for this project were to:

> Use appropriate digital technologies that enhance and augment the personal and embodied experience of viewing the textile AIDS Quilt; raise awareness about the multiple stories of the Quilt panels; raise awareness about the archiving needs for the textile Quilt; communicate the social importance of this work of international cultural heritage; raise awareness about the contemporary status of AIDS/HIV in an international context; promote the Quilt as a living memorial.

The first phase of the project focused on two efforts: (1) to create the backend data architecture, and (2) to clean up and import the large data sets already collected by the NPF. The data sets include the collection of high-resolution images of 5,600 blocks, and accompanying meta-data about individual panels. The outcome of this phase resulted in the creation of a Virtual AIDS Quilt, a 25-gigapixel image comprised of the 5,600 block images digitally stitched together. The NPF only photographs blocks, not individual panels. The 5,600 blocks include approximately 48,000 panels. The Virtual AIDS Quilt digital image arranges the block images according to accession numbers, which roughly corresponds to the historical order of panel acquisition. We know that the first 100 blocks were created from the first batch of panels submitted for the 1987 march on Washington. After that, blocks were assembled as panels were submitted. For most of the thirty years of the AIDS Quilt creation, a woman named Gert McMullin has been solely responsible for bundling individual panels into blocks (Figure 10.5).

Sometimes Gert created blocks out of panels that came in at the same time, other times she waited for a collection of panels to coalesce, which means that individual panels are not necessarily bundled in chronological order. Nonetheless, the Virtual Quilt is a historical document: the earliest blocks show up in the upper left of the image, the most recent blocks are located in the lower right (Figure 10.6). After the research team created the Virtual Quilt image, we turned our attention to creating a browsable interface for the image.

The Virtual Quilt browser enables visitors to explore the Quilt as a spatialized image; in so doing it enacts a "poetics of interactivity" designed to evoke an appreciation of the different scales of significance of the Quilt (Morse 2003). We came to realize that the digital experience could invoke a poetics of perception that was not possible to experience in bodily form. What we could do with the digital, that is not easily accomplished with the textile Quilt, is bridge the scale of viewing experience, from the ground-level experience of viewing the Quilt as one stands next to it, to a bird's-eye view of the Quilt from the perspective of a higher altitude.

The AQT project provides viewers with access to a digital archive of Quilt images as well as to stories about the creation of the Quilt and the rise of

FIGURE 10.5 Image of Gert McMullin sewing a Quilt panel.
Source: Image from archive of NAMES Project Foundation. Used with permission.

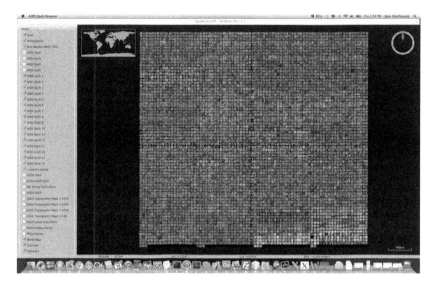

FIGURE 10.6   Image of the Virtual Quilt. AIDS Quilt Touch archive.

HIV/AIDS in the U.S. Like all memorials, the AQT interactives serve as the stage for the manifestation of a broad range of relationships among viewers and technologies. As a media system, it includes elements that are simultaneously cultural and technological: material works as well as digital representations of quilt panels, discursive descriptors (metadata tags) as well as textual accounts (stories, memories, recollections), and unique contributions from individual human agents (quilt makers, activists, health care providers) as well as from social collectives (audiences, families). The media system also includes new practices and protocols, and emergent modes of interactivity.

Our challenge was this: how do we respect and maintain the cultures of the Quilt while developing a digital expression of its essential qualities? We began by asking how the intimacy of seeing the physical Quilt could be matched by digital applications. In designing these applications, we devised methods for visually representing Quilt data sets to enable new insights and the produc- tion of new knowledge. The design process drew insights from the history of public art as well as the histories of public discourse about HIV/AIDS. As works of public art, these applications were created to evoke new perceptions through experiments with scale, mobility, and modes of human engagement in public spaces. As a mode of public communication, these public interactives were designed to engage people in conversation about the impact of the AIDS epidemic, the richness of lives lost, and the contemporary status of AIDS/HIV infection in the U.S. We used the term *public interactives* (Balsamo 2016) to name a genre of interactivity in public spaces that incorporates computationally enabled responsive surfaces and serves as the stage for spontaneous social en- counters. In creating these dynamic media forms, we were especially interested in staging experiences that communicate with younger people who, having been born in the 1990s and 2000s, are growing up in a very different culture than that of the 1980s when the AIDS epidemic began. In this sense, the AQT experiences serve the broader cultural purpose of creating a digital memorial and a contemporary context that can bridge generational interests.

In 2012, we were invited to participate with the NAMES Project Foundation in a large event to display the Quilt once again on the Mall in Washington, DC. During the months of June and July, portions of the Quilt were on display during the Smithsonian Museum's Folk Life Festival and at more than 50 locations throughout the city as part of the NAMES Project Foundation's *Quilt in the Capital* campaign. Working in collaboration with the NAMES Project Foundation, a distributed team of research-designers from the University of Iowa, Brown University, the University of Southern California, and Microsoft Research collaborated to create three *AIDS Quilt Touch* digital experiences: (1) The AQT Virtual Quilt Table-Top Browser, (2) the AQT mobile web app, and (3) the AQT Interactive Timeline.

Working in collaboration with computer graphics pioneer Andy Van Damm, and his computer visualization research team at Brown University, the AQT team created a version of the Virtual Quilt that was comprised of images of the entire Quilt at different levels of resolution. These images were aligned so that it became possible to move along the z-axis of the stacked images. This enables the sense of zooming into and out of the image as if seeing it from different altitudes of viewing. Van Damm's team at Brown had already created an application called the "Touch Art Gallery" (TAG) that enables the viewing of large, digitized artworks through a zooming function that displays the work at different levels of resolution (2014). The first TAG project focused on the Garibaldi Panorama, a multimedia work of art that is 4.75 feet tall and 260 feet wide, painted on both sides. Working with the Virtual AIDS Quilt provided an opportunity for the Brown team to refine the TAG interface with an even larger spatialized artwork. The contribution from Van Damm and his group was especially important for AQT because they solved one of the most vexing technical problems of the project: how to serve a 25 gigapixel image so that the image refresh rate allowed for a smooth and seamless viewing experience.

The Virtual AIDS Quilt was optimized for display on a touch enabled interactive table (Figure 10.7). By allowing for multiple viewing altitudes, the AQT Virtual Quilt Tabletop Browser encourages users to engage with the scale of the Quilt, and to move from a consideration of its immensity and the physical expanse, to a meditation on the affective details stitched into individual panels (Figure 10.8).

The physical size of the table (three feet by four feet), and its horizontal orientation enables multiple people to collaboratively browse the Virtual Quilt. This mode of interactivity re-embodies the experience of exploring a digital archive to make it social and communal. In presenting a close-up view of individual panels, this application showcases the media-rich texture of panels that includes

FIGURE 10.7 Image of visitors using the AIDS Quilt Touch Virtual Quilt Tabletop Browser. AIDS Quilt Touch archive.

FIGURE 10.8 Close up image of the AIDS Quilt Touch Virtual Quilt Tabletop Browser. AIDS Quilt Touch archive.

photographs, decorative stitching, and a range of memorabilia. By enabling people to browse and read individual panels, the application promotes social engagement among multiple users who often collaborate on reading panels together.

Researchers in the field of Human-Computer Interaction make the important point that tabletop computational surfaces actually involve both "interactive" user experiences and "non-interactive" experiences—where people gathered around a surface watch and view other people's actions (O'Hara 2010). We note that when small groups of people encounter the AQT tables, different behaviors emerge. Some people interact with the device, others watch interactions. One of the interesting things we observed was the emergence of a "non-interactive" user behavior that involves shooting digital images of the browser in use. Very often people would annotate their experience of viewing the Virtual Quilt by taking photographs of the digital images displayed on the tabletop (Figure 10.9).

Like creating a rubbing of an etching on a memorial wall or gravestone, photographing the digital image of a Quilt panel functioned as an emblem of witnessing. The desire to capture the act of "digital witnessing" is one of the unexpected outcomes we watched happen time and again.

Collaboration with public artist Jon Winet and his research-design team at the Digital Studio for the Pubic Humanities and Arts at the University of Iowa resulted in the creation of the AQT responsive web app (Figure 10.10).

FIGURE 10.9 Image of visitor taking photograph of panels displayed on AIDS Quilt Touch Virtual Quilt Tabletop Browser. AIDS Quilt Touch archive.

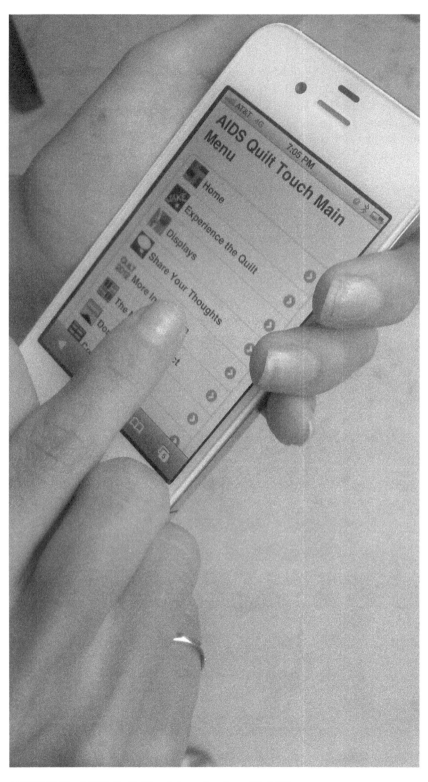

FIGURE 10.10 AIDS Quilt Touch launches June 27, 2012.
Source: Photo by Sherry Moore, 2012. Used with permission.

At its most fundamental, the app enables people to search the database of images of Quilt blocks by name or block number. During the 2012 event, users were also able to find the location of a Quilt block when on display in Washington, DC. The app also provided an opportunity for users to contribute celebrations and comments about the Quilt or individual names. Given the sheer size of the Quilt and the logistical difficulties associated with displaying it, the AIDS Quilt is difficult to keep in the public eye. Online, global, multi-platform access can help keep the Quilt visible.

Comments contributed by members of the public not only addressed personal memories and lost loved ones, but also sometimes spoke about the value of the AQT application. For example, one commenter annotated a panel to document familial relationships: "Mother of Bill and Mark Bolivar. Posted by Ms. Carolee Egan on July 19, 2012." Another narrated the emotional experience of seeing loved ones' panels again after a long gap in time:

> Last night a friend told me about the quilt returning to D.C. I opened this site and was so pleased to see what has been done with the quilt. After typing in my sons' names I watched the panel appear before me and I began to cry uncontrollably. It was last seen by me when it was on the Mall many years ago.

Others expressed gratitude for the act of memorializing their emotional labor in contributing to the quilt:

> Thank you for caring for the memorial my daughters and I made to remember the boys. I am now seventy-four years old and I will be getting into my car today and driving to D.C. to view it once again. It means so much to me to know that everyone can view it on this site so they will never be forgotten.

The AQT web app produces a kinesthetic intimacy with images of the Quilt. As the most broadly accessible experience, it introduces a new mode of annotating the Quilt, by allowing, via the online text-based submission system, for user-generated personal contributions. Enabling users to participate in the process of annotation addresses the public/private dynamic that is at the heart of the Quilt's commemorative and affective powers; these contributions add new layers of cultural significance to the Quilt archive.

In advance of the Quilt in the Capital event, Microsoft Research provided equipment and funding to support the creation of a third AQT application that enables visitors to read key episodes in the twenty-five-year history of the Quilt and the thirty-year history of the AIDS/HIV pandemic (Figure 10.11).

This application was a collaboration of the Public Interactives Research Team at University of Southern California and the Chronozoom Team at Microsoft Research.

FIGURE 10.11 Image of visitor using AIDS Quilt Touch Interactive Timeline.
Source: Photo by Sherry Moore, 2012. Used with permission.

The AQT Timeline was designed to inspire people to have conversations about the broader social, political, and bio-medical events that are part of the multi-faceted history of AIDS and the Quilt in the U.S. In creating inter-active content, the design team took into account the critiques of the way in which AIDS and the understanding of HIV have been narrativized in the "official histories" of the epidemic. Drawing on the history of AIDS activ-ism, this application presents key episodes highlighting critical acts of social intervention—when, for example, activists confronted government representa-tives and protested official policies. The interactive makes it possible to present multiple histories—both those that serve as official accounts and those that serve as counter-narratives. Many scholars and activists contest these histories for the pejorative bias that sneaks into descriptions and accounts. For example, the persistent reference to people who are infected with HIV as "AIDS victims" propagates identities that are not consonant with those promoted by activists and people living with HIV. The term "victim" implies a state of powerless-ness. The lessons to be learned and remembered from these histories spread across the key domains of human life: there are lessons to be learned about the social practices of medicine and epidemiology, about the popular under-standing of sexuality, about the history of art and activism, and about ongoing tensions between theory and practice. As cultural historian Paula Treichler (1999) writes,

> The AIDS epidemic is cultural and linguistic as well as biological and biomedical. To understand the epidemic's history, address its future, and learn its lessons, we must take this assertion seriously. It is through the careful examination of culture that enables us to think carefully about the meanings of the AIDS crisis: to use our intelligence and critical faculties to consider theoretical problems, develop policy, and articulate long-term social needs even as we acknowledge the urgency of the AIDS crisis and try to satisfy its relentless demand for immediate action.
>
> *(1)*

As Treichler argues, the AIDS epidemic showed us the powerful impact of organized activism. Armed with art, theory, and an anger that would not be denied, AIDS activists moved mountains and governments to change medical practices and policies in the U.S. The lesson for media studies students here is simple but profound: theory combined with practice can change the world.

In 2015, we received a second grant from the National Endowment for the Humanities to build a new version of the AQT mobile web app that enables viewers to participate in archiving information and stories about individual Quilt panels. This phase of the project included the participation of a research team from Michigan State University that developed The Quilt Index as an online platform for archiving collections of textile quilts. The focus of this collaboration was to design strategies of public engagement through the creation of incentives and opportunities for "civic archiving." The Quilt Index project archives the AIDS Memorial Quilt as a collection of individual panels; much of their effort focused on creating a searchable database of images of individual panels.

The AQT team pursued a different approach to archiving the AIDS Memorial Quilt that keeps the focus on the Quilt in its entirety, but in a way that enables people to view individual panels as part of the block they are stitched into. The AQT interface continues to enable people to access panels and blocks by searching a database of metadata or images, but now the application includes a "narrative interface" that provides viewers with opportunities to explore particular themes or follow tours through Quilt images. The application incorporates several narrative "threads" that users can select to guide their travels through the Virtual Quilt to explore topics such as "The History of the Quilt" and "The Art of the Quilt." Our aim is to address those visitors who may not know much about the Quilt's history, nor have a personal connection to someone commemorated on the Quilt. In this way, AQT is expanding its mode of engagement to reach across generations, to address not only those who lived through the early devastation, but also those who are still "living in the age of AIDS." The efforts of AQT now include designing strategies for public engagement through the creation of incentives and opportunities for "civic archiving." We are especially interested in connecting to panel makers by providing an application that easily enables them to record and upload media-rich contributions that enhance stories about individual panels. The aim is to motivate members of a "community of interest" to engage as a "community of participants" in archiving the Quilt. In doing so, we seek to connect our current cultural moment with a cultural formation created in a different era.

Lessons Learned

The AQT project unfolded over a long period of time, in the context of different settings and in collaboration with diverse partners. Funding resources and technology support came from several other partners: the National Endowment for the Humanities, various academic institutions, and Microsoft Research. At one point, we created an Indigogo campaign to raise funds to support the installation of the AQT applications at the 2012 Washington, D.C. event (Figure 10.12).

The development of AQT required the expertise of people trained in diverse disciplines. Humanists provided theoretical context and insights about cultural rituals of remembering and witnessing. Public artists organized displays and events. Computer scientists addressed the difficulties in working with large visual data sets. Designers iterated on interface languages and modalities. The distributed AQT team included humanists, engineers, artists, designers, folklorists, health educators, and computer programmers. The collaborations

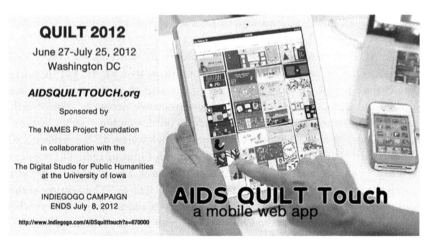

FIGURE 10.12   Image of AIDS Quilt Touch Indigogo campaign. AIDS Quilt Touch archive.

among team members weren't always easy and smooth. There were difficult moments when differences in intent and priorities needed to be negotiated. What was critical to the success of the AQT effort as a design-research project was the willingness of each participant to learn how to navigate differences: in language use, in work practices, in research objectives, in aesthetics, in theoretical viewpoints, and technical expertise. The ability to collaborate across differences is a foundational skill for any person doing applied media studies projects.

The more instrumental lessons learned throughout this project focus on the pragmatic considerations of raising money, supporting research teams, finding equipment, access to a space for production, and managing the development of the project as it takes place in different locations with different people at different times. Funding for applied media studies projects is difficult to find. We adopted a "start-up" mentality by first engaging people in the idea of the project and then doing an inventory of the resources each could contribute. It was important that the project was broad enough so that each collaborator could understand how time invested would pay-off in meaningful ways for them based on disciplinary markers of value and recognition. As the principle investigator (PI) for AQT, I had to learn how to direct the overall project while at the same time listening to the experts involved who often (gently) guided my management and leadership efforts. As a humanist, one of my strong skills is the ability to translate among discourse communities, for example I understand how to parse technical language to make it meaningful to other project participants. On the other side is the skill of translating cultural theory into usability and design language. As a tenured faculty-researcher I enjoyed a measure of freedom in my ability to pursue the project even when it didn't have strong funding. I was able to keep it going, in between the grants, by engaging students, research assistants, like-minded colleagues in what has now become a passion project for many of them. As I've argued elsewhere, the AIDS Memorial Quilt is a boundary object that enables the creation of community at different scales and different temporalities. It is a work of cultural heritage that speaks to people from different vantage points. AQT builds on the cultural significance of the AIDS Quilt, with the aim of expanding that significance through the use of new technologies and new cultural practices of remembering and witnessing.

References

Balsamo, Anne. (2016) "The Cultural Work of Public Interactives," in Christiane Paul (ed.), *The Blackwell Handbook of Digital Art*, New York: Blackwell.

Balsamo, Anne, Dale MacDonald, Nikki Dudley, Kayla Haar, Lauren Haldeman, Mark NeuCollins, Kelly Thompson, Jon Winet and the NAMES Project Foundation. (2012) *AIDS Quilt Touch*. http://aidsquilttouch.org.

Kurin, Richard. (2013) *The Smithsonian's History of America in 101 Objects*, New York: The Penguin Press.

Morse, Margaret. (2003) "The Poetics of Interactivity," in Judy Malloy (ed.), *Women, Art, and Technology*, Cambridge, MA: MIT Press.

O'Hara, Kenton. (2010) "Interactivity and Non-interactivity on Tabletops," *CHI '10: Proceedings of the SIGCHI Conference on Human Factors in Computing Systems*, pp. 2611–4. doi:10.1145/1753326.1753723.

Touch Art Gallery, Brown University. (2014) http://cs.brown.edu/research/ptc/tag/

Treichler, Paula. (1999) *How to Have Theory in an Epidemic: The Cultural Chronicles of AIDS*, Durham, NC: Duke University Press.

Zarembo, Alan. (2006) "The Quilt Fades to Obscurity: Once a Mighty Symbol of Love and Loss, the Tribute to Victims of AIDS Has Gone from Large to Largely Forgotten." *The Los Angeles Times*, June 4, A1. http://articles.latimes.com/2006/jun/04/science/sci-quilt4.

11

THE TIME AND STRUCTURE OF CROSS-COLLEGE COLLABORATION

Developing a Shared Language and Practice

Heidi Rae Cooley

On a fundamental level, urban renewal is grounded in the context of geography. The [Ward One] app approaches this by being heavily location-driven. The interactor is meant to physically explore the former Ward One area, visiting historic sites and seeing what is and isn't standing today. When they begin using the app, they are shown a map of Columbia that displays the boundaries of Ward One and the locations of different sites. This is supported by the rich content that the app provides in the form of text, images, audio, and video.

—*Austin Pahl, Junior, Computer Science*

The [Ward One] app is not intended to be a condemnation of USC or a memorial to Ward One; rather, it intends to educate about the human cost of urban renewal using Ward One as an example and to preserve the history of this neighborhood that would otherwise be forgotten.

—*Bonnie Harris-Lowe, Junior, Media Arts*

Today, we live in a very tumultuous time from black students' [perspectives], and USC is a very tumultuous place... I joined the [Critical Interactives: Ward One II] class because, despite having grown up in South Carolina, I was beginning, for the first time in my life, to understand what blackness was, and I felt like Ward One II could help me explore that. I was starting to understand myself less as a person who happened to be black and a woman, but as a black woman, in a predominantly white institution, with a poor history of handling racial matters. I want the app to be for people like me... But for it to really reach the student population of USC, we have to confront USC's problem with race. The good thing is, I'm starting to think that the app is also a way to start that conversation.

—*Amber Brown, Senior, Media Arts*[1]

My collaborator, computer scientist Duncan Buell, and I team-taught our first cross-college course in fall 2011. Called "Gaming the Humanities," the course drew its inspiration from a three-week NEH-funded summer Humanities Gaming Institute we, along with another colleague, organized and hosted in June 2010. The 2011 course brought together twenty-seven students from computer science, composition and rhetoric, creative writing, media arts, and film and media studies. In tandem with weekly readings that addressed topics such as game design and play, computational thinking, and project development, the students conceptualized and pitched humanities-oriented games. They then formed five interdisciplinary design teams to begin to build some of them. For the most part, the course proved successful; only one design team (the least disciplinarily diverse) struggled to deliver a humanities-based game. Those who attended the end-of-term presentations and demonstrations of prototypes responded enthusiastically. These included, in addition to colleagues and university administrators, representatives from both the local public educational television station and the local independent cinema. Overall, it felt like a successful pedagogical experiment.

   We took away two principal insights from that first foray into collaborative pedagogy, both of which inform how we have approached subsequent iterations of the course. First, we needed to reimagine our own roles if we were to engender a more collaborative atmosphere that would allow for the kind of interdisciplinary project development we hoped to facilitate. "Facilitate" became—and remains—a foundational term for us as we refine our method and practice as collaborator-mentors. We have come to think of ourselves as *project managers*, that is, as members of a larger team who accept responsibility for overseeing the various activities, discussions, and related exchanges that make possible a productive, efficient, and successful design and development process but who do not "own" the process or its outcomes. Our second insight, which is a corollary of the first, was that we needed to restructure the class. We could not function as effective, that is, responsive and responsible project managers if we were overseeing multiple projects; we needed to limit the number of projects. In fall 2012, our course, newly named "Critical Interactives," focused on the development of a location-based proof-of-concept application for iPad, *Ghosts of the Horseshoe* (*Ghosts*), which presents the history of slavery that made physically possible the historic Horseshoe, formerly the grounds of South Carolina College (est. 1801) and currently the "heart" of the University of South Carolina's Columbia campus.[2]

   Subsequent versions of the course (spring 2014, 2015, 2016, and 2017) have allowed us to hone our method, which at a very fundamental level aspires to cultivate habit-change in students, colleagues, and citizens of

Columbia, SC. Broadly speaking, habit-change refers to shifts in how groups of people make sense of things. Unlike feats of persuasion by means of argument or the attempt to change individual minds, habit-change is social and occurs over time. In the context of the Critical Interactives class, habit-change transpires as students wrestle with conventional disciplinary thinking that defines them as majors, serves as a measure of their disciplinary proficiency, and arms them with a sense of confidence (e.g., mastery) that can also—and frequently does—serve to limit or interfere with their ability to communicate with people trained in other fields. The class asks students to develop a new way of thinking and speaking *together*. In other words, we ask students to join us in acquiring a shared language, one that has been evolving since Duncan and I first began collaborating. In what follows, I offer a descriptive account of this process, which has involved colleagues from various disciplines and now includes collaboration with and feedback from Ward One Organization members. Evidence suggests that this process has indeed produced habit-change that is beginning to foster cultural renewal in Columbia, South Carolina.

Evolving and Sustaining a Shared Language: Cultivating a "Community of Interpreters"

Interdisciplinary collaboration requires effective communication. How to accomplish this obvious goal is not so obvious a matter, especially if one endeavors to evolve over the course of several years a sustainable and malleable language that is shared by individuals whose educational backgrounds vary substantially. Duncan Buell and I have been reasonably successful in this regard. In part, this is because we are fortunate enough to have students who enroll in the Critical Interactives course more than once. But also, each semester we craft a syllabus, structure the course, and run class meetings in a manner that invites students to enter into and refine an ongoing conversation. This transpires over the course of the semester and by means of various modes of engagement with readings and previous versions of the project (content and code), as well as through experimentation with user interface and user experience design, which requires thinking carefully about how we understand our audience.

We begin with a conventional seminar style approach. We "curate" readings that serve as points of departure for class discussions. One of the first assigned readings is selected from articles that Duncan and I have co-authored (Buell and Cooley 2012; Cooley and Buell 2012, 2014, 2018). Our rationale for assigning one of our own publications is trifold. First, it provides students with an overview of the project. Second, it introduces them to the vocabulary we use (e.g., "critical interactive," "empathic awareness," "interactor") to discuss the work we do and its aims with respect to the university and the broader community. Third, it demonstrates how a computer scientist and a humanist speak *together* about the projects we develop collaboratively. We also, when possible, assign student authored materials that explicitly address topics relevant to the course.[3] In part, we hope to communicate to students not only that we value student intellectual labor but also that the result of that labor might very well "live on" beyond the limited time of the standard final assignment or exam, or the conferred degree. Additionally, we assign texts and digital based projects that pertain to the particular semester's emphasis (e.g., slavery and universities, urban renewal, oral histories).

Our goal is to scaffold students' entry into active cross-disciplinary conversation that readily translates into doing and making. While we might start any particular discussion seated in a circular formation, we frequently transition into active mappings of central concepts, key words, themes, etc. Quite literally, on white/chalkboard or poster-sized post-its, we draw lines of connection between keywords and phrases, talking through how and why certain relations are more obvious or subtle, more or less significant or central than others to our continuing conversation (Figures 11.1 and 11.2). The resulting diagrams represent degrees, hierarchies, and trajectories of relation that also reveal "sites" of contestation or disjunction, and open onto further discussion and negotiation. Sometimes the process happens in small interdisciplinary groups, where students grapple with an isolated set of concepts (Figures 11.3 through 11.5). Later, we will use this same process to discuss matters of design and refinement of user interface (UI) and user experience (UX), and to consider how we think about "users" (whom we have come to refer to instead as "interactors") (Figures 11.6 through 11.8). We find these humanistic modes of thinking useful because they are recognizable to computer scientists who regularly brain-storm and create wire-frame maps.

Humanists get a similarly friendly process-oriented introduction to programming as agile development. A volunteer advanced computer scientist leads the class through several programming exercises (e.g., "hello world," building a clickable button, changing font color and size, etc.); the other computer scientists, including Duncan, sit beside the humanists—like me— and serve as guides. Humanists quickly learn how unforgiving code can be: a rogue brace or a mistyped class name, for example, results in software that just won't run. The exercise also underscores the quantity of time and the kind of attention to detail that programmers invest in writing code and, therefore, emphasizes the absolute necessity for timely delivery of humanist-produced content.

During the course of a semester, students become increasingly engaged with—and committed to—the project and its development. Their discussions

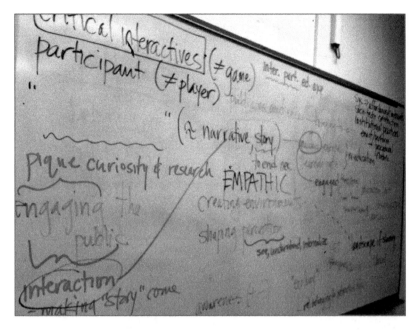

FIGURE 11.1 Concept Mapping (Spring 2014).

Source: Photo by Heidi Rae Cooley.

FIGURE 11.2 Brainstorming Content and Goals (Spring 2016).

Source: Photo by Heidi Rae Cooley.

FIGURE 11.3 Group Work—Context (Spring 2015).
Source: Photo by Heidi Rae Cooley.

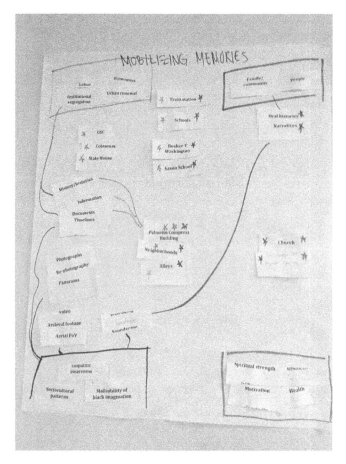

FIGURE 11.4 Group Diagrams (Spring 2015).
Source: Photo by Heidi Rae Cooley.

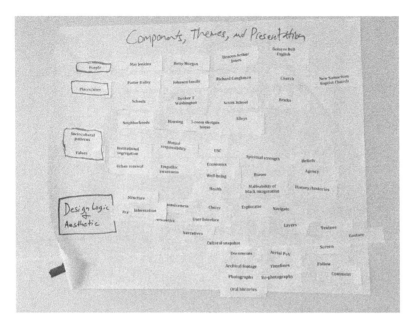

FIGURE 11.5 Group Diagrams (Spring 2015).
Source: Photo by Heidi Rae Cooley.

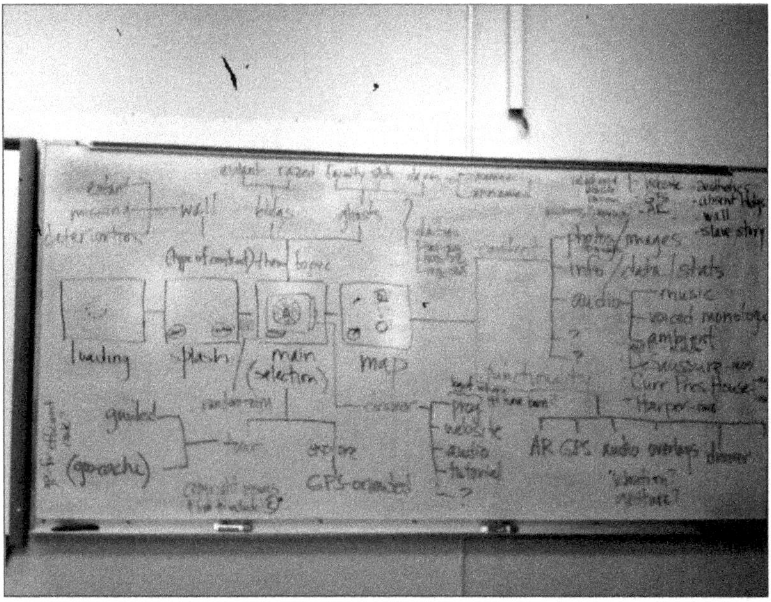

FIGURE 11.6 User Experience Diagram—*Ghosts of the Horseshoe* (Fall 2012).
Source: Photo by Heidi Rae Cooley.

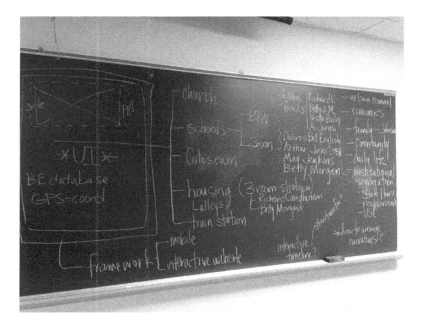

FIGURE 11.7 User Experience Diagram—Ward One app (Spring 2015).
Source: Photo by Heidi Rae Cooley.

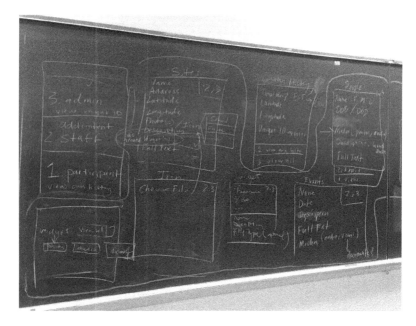

FIGURE 11.8 Student-generated Wireframe for the Ward One App (Spring 2016).
Source: Photo by Heidi Rae Cooley.

about content and its method of delivery to the mobile touchscreen device communicate a shared concern for and responsibility to the histories being presented. For example, in the spring 2016 iteration of the course, students interacted with and recorded the personal stories of individuals who had lived as children in Ward One in the 1950s and 1960s. They took great care to ensure that each individual who shared a story had his or her voice included in the app. Both in class and in team-messaging exchanges (Figure 11.9), student discussions about content, app functionality, and interface design readily crossed disciplinary domains: computer scientists edited moving image materials to produce thematic compilations of archival footage and interviews; humanists actively contributed to conceptualizing the user interface for uploading assets to the content management system. Students applauded individual and team accomplishments and respectfully pushed their colleagues when the rate of production proved slower than desired. And they navigated frustrations tactfully when others failed to meet deadlines. One of the lessons they learned, in this regard, was that assuming responsibility for the completion of a task to which someone else had committed him/herself is an unfortunate—and hopefully infrequent—aspect of collaboration.

Time and again, students who enroll in the Critical Interactives courses exhibit a shared appreciation for the unacknowledged histories they endeavor to present, respect for each other's contributions, and a striking degree of poise in speaking about the work they—as a collective—have accomplished.

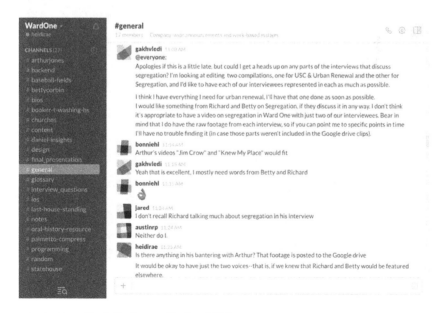

FIGURE 11.9   Slack Exchange (Spring 2016).
Source: Photo by Heidi Rae Cooley.

These qualities take time to develop; they must be cultivated. Over the course of fourteen weeks, Duncan and I challenge students to own and speak about the projects they are developing, and we ask them to keep in mind that their efforts belong to a "legacy" of student labor. The "test" of this process takes place in the culminating event scheduled for the last week of the term, a public presentation and demonstration of the app to an audience of colleagues and other invited guests. Duncan and I participate very little in this event; we only offer a few opening remarks of welcome and thanks. Students perform with eloquence and they respond confidently and articulately to questions about, for example, the development process, motivation for design decisions, and their vision for moving forward. But what most delights and inspires me is seeing them on stage together, sharing the microphone, nodding in affirmation of individual remarks, and speaking in terms of "we" (Figure 11.10). In moments of impressive intellectual generosity, they move beyond merely being peers; they are colleagues, collaborators, co-conspirators: they evolve into what in Peircean terms is a "community of interpreters" (Anderson and Hausman 2012: 139).

FIGURE 11.10   End-of-term Presentation and Demonstration—Question and Answer (Spring 2016).
Source: Photo by Heidi Rae Cooley.

Expanding a "Community of Interpreters": Bringing Faculty Experts into the Conversation

For Peirce, the interpretive process is a process of inquiry. It is inherently social, belonging to a community that is always potentially expanding. The process is regulated by norms—by which Peirce means shared habits of thinking and practice—that shape opinion and corresponding action. At the same time, inquiry transpires according to the general (i.e., common) "would be" of the future conditional tense and, therefore, is not guaranteed. Any agreement or consensus in interpretation lay "in the manner in which [individual] reactions contribute to that development [of consensus]" (Peirce 1974: 2 [5.3]). Together, individuals strive toward reaching agreement about the questions and concepts that inform the interpretive pursuit. In other words, it's not about any particular answer as such. In the case of the Critical Interactives class, the Ward One app, as a discrete instance of functioning software, is not the point; rather, the app provides the means for having conversations about race in the context of the USC, Columbia campus, and the history of racial politics that made possible the physical plant of the institution and its continued expansion. Facilitating productive conversations about these issues, which are beyond the scope of Duncan's and my respective disciplinary training, has meant that we have turned to colleagues for assistance and insight.

The most straightforward example of how we bring faculty from various disciplines into the conversation is the mid-term presentation. Each version of the Critical Interactives class requires students to present work-in-progress at the midpoint of the semester. To this event, we invite faculty whose areas of expertise pertain to the questions the class is pursuing. These have included questions about the hiring out system of slavery, urban renewal, and racial representation, how to write for the public, effective interface design, and use of archival, still and moving image, and audio materials.[4] At the conclusion of the mid-term presentation, the faculty audience asks questions and offers critical commentary. In spring 2016, much of the post-presentation discussion centered on the selection and inclusion of the voices and stories of former Ward One residents, people who were children and teenagers at the time of the University's acquisition of Ward One properties. In this context, architectural art historian Lydia Brandt and film and media historian Mark Cooper urged the students, who had begun meeting informally with former residents and were still preparing for more formal interviews, to find ways to move conversations beyond simple recollections of place in the past. They encouraged them to draw out the personalities of their interviewees in order to have a multidimensional sense of them as individuals who have opinions, frustrations, hobbies, and who have experienced life beyond Ward One. Here, they were pushing students to conceptualize more complex narrative structures that the app might employ to underscore tensions, continuities, and discontinuities between past and present, individuals and community, and the university and the local surroundings. Moreover, they prompted students to directly ask former residents about their feelings about USC, both past and present—something that historian Bobby Donaldson had suggested earlier in the semester.

Students are in conversation with faculty and other university experts much earlier in the semester than the mid-term presentation of work-in-progress, and they maintain these relationships after the mid-term presentation as circumstances require. When the Critical Interactives class was developing *Ghosts of the Horseshoe*, we asked historian Robert Weyeneth to introduce the students to the *Slavery at South Carolina College, 1801–1865* website. Once the class focused its efforts on the Ward One app, Bobby Donaldson attended one of the first class meetings to discuss the Ward One community, share archival materials and personal accounts about former Ward One residents, and provide an initial overview of urban renewal in Columbia, SC. In addition to colleagues, we solicit the assistance of faculty librarians. For the *Ghosts* project, University Archivist Elizabeth West gave tours of the Horseshoe campus, one of which focused on the historic wall (erected in 1835–1836) that encloses the historic campus. For the Ward One project, we have secured the assistance of the University Oral Historian Andrea L'Hommedieu. In spring 2016, she offered a workshop on best practices for conducting oral histories, including how to develop open-ended questions, where and how to record interviews, and the importance of getting signed release forms.

These experiences with university faculty and faculty librarians, as well as other university professionals, expand students' capacity to think about the cultural and institutional value of preserving, on one hand, seemingly banal, utilitarian structures such as the brick wall and, on the other hand, the stories of those for whom access to university grounds, let alone classrooms and other facilities (e.g., libraries), was strictly prohibited. In acquiring a new vocabulary for speaking about how the *place* of the university matters, students acquire the ability to intervene in what and how the institution means to those who find themselves on or near campus. Moreover, this vocabulary, and the interpretative process by means of which it expands, makes possible serious discussions about issues and questions that intersect with those taken up by the critical interactive under development. A recent conversation with University Architect Derek Gruner, who attended the end-of-term presentation and demonstration of the 2016 prototype of the Ward One app, provides a poignant example of this. Gruner recounted an exchange he had with a couple of students at the conclusion of the event, after the formal question and answer session. He explained that he had presented them with a question that he has been grappling for several years: Should the Coliseum, formerly a large multi-purpose arena, be razed? The students' collective response did not really surprise him. For them, the structure, built in 1968 on grounds acquired by the university as a result of Columbia's "Fight the Blight" campaign, stands a reminder of the kinds of hostile, racially-inflected appropriation that urban renewal initiatives made possible across the nation; they had no problem with it being slated for demolition. Gruner then asked whether or not it matters that the Coliseum is architecturally significant as a unique example of mid-century modern brutalist architecture; they were less confident, but remained committed to the possibility of demolition. The fate of the Coliseum remains unclear (although preservation is cost-prohibitive); but this account is significant because it shows how capacious the "community of interpreters" can be. That there is room for diverse opinions, for instances of disagreement, makes possible more complicated discussions with a variety of potentially interested parties.

Reaching Out to the Community: Collaborating with Ward One Organization Members

Duncan and I have always intended the work of developing critical interactives to be about reaching out to a broader community. Beyond the students who enroll in the class and become stewards of its ongoing development, we have considered it crucial that the general student population, comprising many ardent supporters of the school, should be aware of the complicated histories manifest in the university campus they daily inhabit.[5] We are not interested in cultivating guilt for the histories that precede the current generation, but we are invested in how students might begin to see themselves as agents of change. We would like for them to recognize the lasting effects of years of racial inequity—effects that mark the physical campus as well as shape the university's student body (both demographics and attitudes)—and, subsequently, *reconsider* their relationship to that past as they themselves live, study, and work in the present. Moreover, we imagine that visitors to campus and local residents might also develop a greater appreciation for how race has shaped the contemporary landscape and interpretations of the university as a place. In grant prose, we have used the phrase "cultural renewal" to describe what we hope to accomplish by building the Ward One app. More recently, we have borrowed a term from digital media artist Sharon Daniel (2016)—"reconsideration"—in order to underscore that we propose neither to resolve issues nor to make accusations pertaining to past or current injustices. Instead, we hope to inspire dialogue generative of new connections, new insights, new ways of thinking about the university that might motivate the community to imagine a shared and equitable future on and around campus.

With the help of Bobby Donaldson, we have reached out to the Ward One Organization, many of whose members were residents of the old ward. In spite of an initial period of trepidation, Ward One Organization members have come to embrace the project; they have begun to entrust us with their stories. While students convened with Ward One Organization members in spring 2015, they did not begin conducting interviews until spring 2016 (Figure 11.11). The months in between provided Duncan and me with opportunities to learn more about the Ward One Organization and its membership, and time to approach our colleague, documentary filmmaker Laura Kissel. She agreed to coordinate with her documentary production class to film interviews with former Ward One residents. On a Saturday in January 2016, students from both the Critical Interactives class and the documentary production class met with a dozen members of the Ward One Organization in the Booker T. Washington Auditorium, the venue where nearly all Ward One representatives present had graduated high school (when the facility was known as Booker T. Washington High School). Both Donaldson and Kissel joined us, Donaldson frequently guiding the conversation, prompting former residents with archival photographs he projected onto the screen. After introductions, individual Ward One members took turns sharing stories about life in the old ward; students asked follow-up questions. The conversation lasted for two and a half hours (Figure 11.12).

In our next class meeting, student design-production teams identified the individuals they most wanted to interview. With Ward One Organization president Mattie Roberson's assistance, we contacted potential interviewees. By mid-February, teams of students from both the Critical Interactives class and the documentary production class began meeting informally with the five individuals who accepted invitations to participate in the interview process. Based on those conversations, and guided by principles of oral history as outlined by L'Hommedieu and supplementary class readings, the students developed

FIGURE 11.11 Meeting with Ward One Organization and speaker Mattie Roberson (Spring 2015).

Source: Photo by Heidi Rae Cooley.

FIGURE 11.12 Meeting with Ward One Organization (Spring 2016).
Source: Photo by Heidi Rae Cooley.

a framework for the formal interviews. They identified central themes (e.g., community, family, education, gendered labor, free time), and they drafted questions. They refined those questions with Donaldson's help. Exchanges between interview team leads and individual interviewees continued as the official list of interview questions became finalized and times for the formal interviews were scheduled. On Friday, March 18, Kissel led a mix of students in arranging the space for the subsequent day's shoot. We hauled sound blankets and sound kits, light kits, scrims, boom and lavaliere microphones, extension cords, fresh batteries, and other supplies to the Mann-Simmons House (a local house museum) whose "living room" we converted for the purpose of filming the interviews (Figure 11.13). Early Saturday morning, the filming began. In-dividual teams arrived early and greeted their interviewee as he or she entered the door. The students and interviewees engaged in light conversation as Kissel secured their microphones. The students guided their interlocutors to the seat from which they would share their stories. As each individual spoke, a strange quiet filled the room; a keen awareness of the microphones, to be sure, but also a sense of expectation and an eagerness to listen, to understand.

FIGURE 11.13 Setting Up for Filming Formal Interview with Arthur Jones and Documentary Filmmaker Laura Kissel (Spring 2016).

Source: Photo by Heidi Rae Cooley.

FIGURE 11.14   Critical Interactive Students Travis Casey and Bonnie Harris-Lowe Prepare to Interview Former Ward One Resident Arthur Jones (Spring 2016).

Source: Photo by Heidi Rae Cooley.

The voices of Betty Corbin, Arthur Jones, Richard Caughman, and Mattie Roberson compelled all who attended the shoot that Saturday (Figure 11.14). Recollections of churning butter, of cornbread biscuits and scaling fish, of chess, and of not being permitted to walk on certain sides of streets; their stories resonated. In fact, a group of students in the documentary production class decided to base their final film project on Ward One, the former residents, and the Critical Interactives class. The documentary, *Ward One: Reconstructing Memory* (2016), presents an example of what the Ward One app aspires to elicit in its interactors: the "ah-ha" of connection—between a here-and-now and the past as conveyed through the language of those who lived back then. Near the film's conclusion, we hear the voice of Mattie Roberson; she's describing the location of her first job, searching for the name of the place: "it's a tall, real tall; it's residential, and then it has a—" Mattie breaks off mid-description, remembering the name: "Cornell, Cornell Arms." By the time she has arrived at the word "Arms," the students have recognized the place and chime in eagerly: "Oh, yeah, Cornell Arms." Mattie continues, explaining that she used to make pimento cheese sandwiches. One of the filmmakers, still off camera, interjects, "I work in Cornell Arms; in the bottom floor, there's a café." The exchange continues between on and off camera spaces, as the student filmmaker and Mattie realize that they share a connection: the Cornell Arms Café where Mattie once worked is now Carolina Café where the student now works. The moment of discovery is infectious, resulting in the camera panning left to include the student (and two others) in frame, at which point Muriel Lee, the student, expresses with delight, "I make pimento cheese sandwiches, too!" Out of frame, Mattie gasps in surprise and claps her hands. The camera returns its focus on Mattie, her hands clasped; she exclaims incredulously, "So they're still makin' the pimento cheese?" Muriel confirms that, indeed, they are still making the pimento cheese.

The sequence is full of the vitality brought by the unexpected. A place familiar to locals resonates with new associations. Viewers are likely to recall this small but potentially significant realization about pimento cheese sandwiches past and present the next time they pass the Carolina Café. Moreover, the richness of the connection is felt by all present at the moment it occurs, and extends beyond that moment's documentation; we, in the audience, also share in the delight of the pimento cheese revelation. Subsequently, our language for thinking about a particular place, a café located at the corner of Pickens and Sumter Streets, now includes the voices of Mattie Roberson and Muriel Lee. It also includes the knowledge that, while the restaurant changed hands at some point during the forty plus years separating the two individuals' stints behind its counters, pimento cheese sandwiches abide. Moreover, we might appreciate the fact that food service remains a viable form of employment for young adults, even as we might wonder about how the labor of such employment and its compensation have changed. The coincidence of "the pimento cheese" opened onto an "intellectual sympathy" (*a la* Peirce)—or what I call "empathic awareness"—that mediated across a constellation of differences, including age, race, gender, disciplinary knowledge, and life experience. This is the kind of experience that we, in teaching the Critical Interactives class each year, hope to make possible for others—potential interactors—who routinely have in hand geolocative, networked touchscreen devices.

Notes

1 The first two epigraphs are excerpts from student responses to the spring 2016 end-of-semester self-assessment assignment that concludes the Critical Interactives course each term. The third epigraph is from an email to the author (September 4, 2016). I thank the students who granted me permission to quote their statements about the Ward One application.

2 *Ghosts of the Horseshoe* draws upon the work of graduate students enrolled in historian Robert Weyeneth's spring 2011 public history seminar on historic site interpretation, who produced a robust scholarly website, *Slavery at South Carolina College, 1801–1865: The Foundations of the University of South Carolina.* Three separate team-taught courses are responsible for development of the proof of concept prototype of the *Ghosts* app, video capture of which can be viewed at http://calliope.cse.sc.edu/ghosts/. Richard Walker was the lead computer scientist for the app, and the 1.0 version of the software was the core of his PhD dissertation (2014).

3 In the case of the *Ghosts* project (fall 2012, spring 2014), students explored the *Slavery at South Carolina* website (see footnote 2). For the Ward One project (spring 2015, 2016, and 2017), students read two Masters theses by public historians (Richey 2004; Bouknight 2010), whose topics focused on urban renewal and Columbia, SC.

4 Faculty who have attended these mid-term presentations come from the fields of art history, computer science, English, film and media studies, history and public history, and media arts. To date, they include: Jenay Beer, Lydia Brandt, Mark Cooper, Susan Courtney, Bobby Donaldson, Chris Holcolm, Julie Hubbert, Laura Kissel, Allison Marsh, Simon Tarr, and Robert Weyeneth.

5 *Ghosts of the Horseshoe* has been used in the university's introductory freshman seminar, "The Student in the University" (UNIV 101). Duncan and I regularly demo
Ghosts and/or the Ward One app in a variety of classes, including public history,
film and media studies, media arts, computer science classes.

References

Anderson, Douglas R., and Carl R. Hausman. (2012) *Conversations on Peirce: Reals and
Ideals*, New York: Fordham University Press.
Bouknight, Ashley. (2010) "'Casualty of Progress': The Ward One Community and
Urban Renewal, Columbia, South Carolina, 1964–1974," MA thesis, University of
South Carolina.
Buell, Duncan A., and Heidi Rae Cooley. (2012) "Critical Interactives: Improving
Public Understanding of Institutional Policy," *Bulletin of Science, Technology & Society*
32: 489–96.
Cooley, Heidi Rae, and Duncan A. Buell. (2018) "Building Humanities Software that
Matters: The Case of *Ward One* Mobile App," in Jentery Sayers (ed.), *Making Things
and Drawing Boundaries: Experiments in the Digital Humanities*, Minneapolis, MN:
University of Minnesota Press.
———. (2012) "'Critical Interactives': On the Origin of a Concept," in Crystle Martin,
Amanda Ochsner, and Kurt Squire (eds.), *Proceedings GLS 8.0 Games + Learning +
Society Conference*, Pittsburgh, PA: ETC Press, 391–6.
———. (2014) "*Ghosts of the Horseshoe*, a Mobile Application: Fostering a New Habit of
Thinking about the History of University of South Carolina's Historic Horseshoe,"
Annual Review of Cultural Heritage Informatics 1: 193–212.
Cooley, Heidi Rae, Duncan A. Buell, and Richard Walker. (2014) "From Ghosts of the
Horseshoe to Ward One: Critical Interactives for Inviting Social Engagement with
Instances of Historical Erasure (Columbia, South Carolina)," in Hudson Moura,
Ricardo Sternberg, Regina Cunha, Cecilia Queiroz, and Martin Zeilinger (eds.),
*Proceedings of the Interactive Narratives, New Media & Social Engagement International
Conference*, Toronto, Canada, October 23–25, 47–53.
Daniel, Sharon. (2016) "Participatory Media and Social Justice," Artist Talk at
University of South Carolina, Columbia, South Carolina, February 12.
Peirce, Charles Sanders. (1974) [1934/1935] *Collected Papers*, vol. V and VI, (eds.),
Charles Hartshorne and Paul Weiss, Cambridge, MA: The Belknap Press of Harvard
University Press.
Richey, Staci. (2004) "Variation on a Theme: Planning for the Elimination of Black
Neighborhoods in Downtown Columbia, South Carolina, 1905–1970," MA thesis,
University of South Carolina.
"Slavery at South Carolina College, 1801–1865: The Foundations of the University
of South Carolina," University of South Carolina Library Website, Published May
2011. http://library.sc.edu/digital/slaveryscc/.
Walker, Richard. (2014) "*Ghosts of the Horseshoe*: A Mobilization of a Critical Interactive,"
PhD diss., University of South Carolina.
Ward One: Reconstructing Memory (2016) Directed by Emily Acerra, Chandler Green,
Muriel Lee, and Christine Shestko. http://wardonefilm.wixsite.com/memory.

PART V
Intervention

12

UNINTENDED CONSEQUENCES

Kirsten Ostherr, Anne Balsamo, Jason Farman, Elizabeth Losh, Patrick Vonderau, Heidi Rae Cooley, Eric Hoyt, Tara McPherson, and Bo Reimer

Kirsten Ostherr

This chapter addresses the complex, sometimes unintended—but also potentially very rewarding—consequences of intervening into practices that are more commonly studied from a distance, and the political and ethical implications of this work. The case studies for this chapter explore what happens when media scholars become actively involved in the reshaping of media experiences and infrastructures, and in some sense become part of the very processes they seek to critique. Contributors were asked, has your applied media studies work ever produced truly unexpected results that raised unintended ethical or legal issues you had to address? How have you managed issues relating to Intellectual Property regulations with open, collaborative work online? Have you discovered any novel ethical challenges or responsibilities from putting work online—including objects created by people, living and deceased, from cultures that are different than your own? Have you ever been unexpectedly drawn into an applied media studies project as a participant in ways that redefined your understanding of your role as scholar, or expert, or lead investigator? How did you respond, and has this changed the way you work now?

Surprisingly few contributors encountered difficulties relating to Intellectual Property, perhaps due to the fact that none of these enterprises aim to commercialize and thereby profit from their efforts. Nonetheless, almost all of the projects do intervene in some way in the representation of communities in which they do not claim membership. Several respondents describe how they have handled the challenges that inevitably arise once media—especially participatory media—emerge from the development phase and open up to the public at large. All of the contributors acknowledge that the public interface is absolutely essential to their applied media studies projects; indeed, it is often the site of greatest impact and the source of valuable new insights. Many also note the risks that accompany experiments "in the wild," whose unforeseen effects can have both positive and negative impacts on the researchers and the communities of interest.

Has Your Applied Media Studies Work Ever Produced Truly Unexpected Results That Raised Unintended Ethical or Legal Issues You Had to Address? How Have You Managed Issues Relating to Intellectual Property Regulations with Open, Collaborative Work Online?

Anne Balsamo

Several times during the development of the AIDS Quilt Touch project the design team had to confront significant ethical issues pertaining to privacy concerns, internet trolling realities, and the limitations of "crowd sourcing" archival information. Many original Quilt panel makers requested that certain information (last names, city location) about their loved one, or themselves, not be made public until after the death of the panel maker. Even when we knew the panel maker was no longer alive, we had to decide whether to make the information part of the publicly available interactive experience.

We are fortunate to work with the NAMES Project Foundation who provided access to the multi-modal data sets of the AIDS Quilt. There are not intellectual property rights for these materials. The use of the data is governed by privacy policies established by the NAMES Project to ensure panel makers, confidentiality. Issues of intellectual property for the development of the database architecture for the AIDS Quilt Touch project were negotiated as an agreement between the institutions that were part of the NEH grant.

Jason Farman

In a section of one of my classes, I have the students engage with ideas around e-waste and planned obsolescence. For this section I pair sets of readings in the field of repair and maintenance, as well as works relating to e-waste and planned obsolescence, with hands-on exploration of repair. I begin by having my students work with a set of broken mobile devices to replace their cracked screens. We discuss issues around the repair of hardware, such as questions of maintenance and planned obsolescence relating to the operating systems and data bases that make these devices function. Each year, one of the things that we confront in this section is the ethics of repair and how such knowledge should impact my students' everyday lives. For example, I tell them that I bought the mobile phones we are repairing for $10 each, but the digitizers that we are replacing cost $45 each. Then I tell my students that they can purchase this exact phone brand new for $40. Thus, it cost more to repair the phone then it would to purchase a new one. We discuss the larger implications of repairing versus discarding the device, coupled with dialogue about planned obsolescence and how long the device will remain functional as operating systems are updated. My students are torn on the ethics of repair when the act of repairing costs more than simply buying a new device. One student commented,

> Well, now I know about how these devices are produced using unfair labor practices and models of planned obsolescence. But I'm still going to go buy the new iPhone when it comes out. Now I will just feel guilty when I buy it.

From my work in the classroom, I've had to make sure that my students feel a sense of agency with regard to the impact of their consumer decisions. Many feel overwhelmed by the scale of e-waste and obsolescence. They feel that their individual acts of consumerism have very little ability to make a change in this landscape. When they learn that 426,000 mobile devices are discarded each day in the U.S. alone, their decision to hold onto a mobile device for an extra year instead of upgrading seems insignificant. This, coupled with the desire for new technologies that is embedded in contemporary consumer culture, creates a circumstance that does not offer a clear route for my students to pursue. While these practices of applied media studies lead many to change their relationships to their technologies and to "upgrade culture," some students end up feeling resentment toward the kind of knowledge uncovered by this work. Parallel to this, some students study marketing and business and hope to employ planned obsolescence in order to stimulate consumer purchasing for their future companies. While I had hoped to present planned obsolescence as a warning for contemporary society, especially one that is characterized by a ubiquity of digital devices, these students instead took it as a lesson for good business practices!

Elizabeth Losh

In the relatively privileged position of a white straight tenured faculty member, I try to be mindful of the risks of appropriating the digital stories of others. I find Moya Bailey's work on questions of community informed consent really helpful here, along with Amy Earhart's work on how communities have good reasons for mistrust of universities. Looking back, some of the work I did about coming out videos on YouTube created by teens seems problematic in this regard. Recently I have struggled with how to write about the use of Facebook Live in conjunction with the Black Lives Matter movement. When you are talking about digital archives and practices created by people in conditions of precarity, you don't want to make them more vulnerable, and you don't want to oversimplify their digital work as rhetorical actors, since "victim" or "survivor" labels are often inadequate.

Regarding Intellectual Property issues, I generally like to publish in open access environments and to share PDFs of my articles widely. Because I collaborate with people who may be working with students outside the context of the Global North, I don't want to perpetuate barriers to access, particularly when—as John Willinsky (2005) has pointed out—certain academic publishing cartels do depend on free labor from faculty and coerced subscriptions from professional organizations. But it is important to be respectful of those who depend on writing and other creative compositions not only as an income stream but as a way to express very personal anxieties and desires. For example, the community educators working for the Center for Solutions to Online Violence justifiably expect to be paid for writing lesson plans, graphic novels, and other materials for young people to use.

Patrick Vonderau

For me, the very reason for doing this kind of work is of course that I hope to come across something unexpected. In this sense, unexpected results are an expected outcome. This includes legal and ethical challenges. When we applied for the Spotify project at the Swedish Research Council (Vetenskapsrådet), potential legal and ethical challenges were clearly flagged, but not considered a hindrance by the Council. Note that this application explicitly was about establishing fake record labels to disseminate sampled sounds via Spotify, in order to "follow the data" through digital networks. What if these sounds actually turn out to be popular, generating revenue? We also employ a large number of research bots, or fake listeners—it is surprisingly easy to mess around with Spotify's pro rata revenue share model. Our strategy here is to make these findings public, and to use them in order to engage Spotify in a public conversation about its own ethical standards.

In Europe, there seems to be more leeway when it comes to Intellectual Property regulations related to academic work. Seen from here, the U.S. is much more litigious. In many European countries, such as Germany or Sweden, you have the right to quote IP protected materials in argumentative, scholarly contexts.

Heidi Rae Cooley

My collaborator Duncan Buell and I are very aware that the work we and our students are pursuing explicitly draws attention to racial politics and policies that served to establish the university (c. 1801) and make possible its physical formation and expansion—at the expense of black bodies and, later, black families, neighborhoods, churches, schools, and businesses. We are also aware that while the institution has quietly acknowledged its history of slavery, it has been more reluctant to address how it benefitted—and continues to benefit—from federal legislation that supported urban renewal programs subsequently responsible for razing minority neighborhoods across the nation in the mid-twentieth century. So presenting these histories without seeming shrill or unappreciative of our employer has been important. At the same time, we are adamant that these histories and the related personal stories that contextualize them be told as candidly as possible. So, for example, we directly ask former Ward One residents about how they felt about the university as youngsters growing up in its shadows, when, as they describe, they were not allowed to play on the campus grounds, when they were told by their parents they had to imagine going to one of Columbia's HBCUs located across town. We feel fortunate to have navigated this well enough so far. Our work was recognized as providing an example of innovative pedagogy in May 2016 when we were invited to participate in the poster session for the Provost's Summit on Undergraduate Education.

We have not had Intellectual Property issues arise, primarily because neither *Ghosts of the Horseshoe* nor the Ward One app is "live" yet (we're still in the proof of concept-prototype stages in each case). But even so, the archival materials we are using thus far belong to the University's various collections, the already existing scholarship we draw on (especially in the case of *Ghosts*) has already secured the requisite permissions (or examples for how to proceed), and we are careful to acquire completed release forms from our interviewees. Code, user navigation (UX) features, and user interface (UI) assets are either produced by the students or draw on Open Source libraries. All members of each year's design teams—students, colleagues, and affiliated consultants—are credited, as are the Ward One Organization members who have contributed their stories (in the case of the Ward One app).

Eric Hoyt

The Media History Digital Library primarily works with public domain material. Consequently, we can digitize these out-of-copyright works in their entirety and offer broad access to all users. David Pierce, founder and director of the MHDL, has thirty years of experience investigating the copyright status of books and films. Before we scan anything, David researches the copyright status. The key question—at least for works published prior to 1964—is whether or not the rights holder chose to renew the copyright after the initial twenty-eight years of protection.

If you are looking for guidance, there are online resources, such as the University of Pennsylvania's "Online Books Page" and Columbia University's Copyright Advisory Office, that contain lists of public domain works. As Columbia's Copyright Advisory Office warns, though, the "list is not meant to be construed as a silver bullet to completely avoiding copyright" and the University does not make any "warranty that the materials are, in fact, without legal protection." If you are considering spending several thousand dollars on digitization and website development, then hiring a copyright consultant for a couple of hours of work could be a worthwhile investment.

Tara McPherson

I was at an international conference when I got an urgent message from our Project Manager, Curtis Fletcher. He had received a series of angry emails overnight from a group of young women who had discovered a public Scalar project online that reported on and analyzed a closed community in which they participated. Scalar is open software that anyone can use to create a project. We do not know the content of the vast majority of Scalar projects, just as Word-Press or YouTube don't monitor every page created through their platforms. The young women who contacted us are part of a private online group that is "pro anorexia." The author of the Scalar project had joined their group and then written about them without their knowledge or permission.

We contacted the author, who quickly responded that she had created the project for a graduate class and was fine with having it taken down. Had she not responded, we would have taken the project down anyway, as it violated our terms of service. Nonetheless, it led to some serious discussions among our team. As providers of a platform (and one based at a university), what are our ethical responsibilities vis-à-vis how Scalar gets used? What might we do if someone used Scalar to circulate obscene images or hate speech? The legal and ethical implications of such questions do not always neatly line up, although we have tried to address such issues in our terms of service. We are seeing such issues play out on a larger scale as platforms like Twitter and Facebook attempt to address harassment and the spread of fake news. The answers are not cut and dry. For now, we continue to address these issues as they arise and by making sure our terms of service set clear guidelines.

Vectors was launched as a freely available, open access journal in 2005. We were and are very committed to exploring different pathways through which academic work can circulate outside of the tight control of for-profit companies that harvest academic labor and return very little to scholarly communities. (I would count Routledge in that group!) We asked our authors to select a CC license for their work.

In developing Scalar, we knew we wanted to build open source software that could be shared and commonly developed. Our code base is on GitHub and can be installed on other servers and modified. Most Scalar projects are freely available, although we are also working with individuals and with university presses to develop models for selling Scalar projects. One interesting technical aspect of Scalar concerns how we connect to media files. While the media files seem to be "within" a Scalar project, they actually reside in separate archives and servers. This makes it easier for presses to publish work that includes commercial media because the media is not technically in the Scalar project. It resides elsewhere. This feature was incorporated to make it easier for scholars and teachers to work with media without costly permissions and in the spirit of fair use. We also encourage scholars to host their media in Critical Commons, a fair use media archive spearheaded by Steve Anderson.

Bo Reimer

Much of the work we carry out is based on interventions in the city of Malmö. Since this is work carried out "in the wild," almost by definition things will happen that are difficult to predict beforehand. So you know unexpected things will happen, and what you need to do is to prepare for this the best you can. With the participatory design approach that we use, and with the creation of Living Labs, many different stakeholders—with different interests—are involved, and it is important to try to find a way for these to communicate. It is also important to remember that in these situations, the playing field is not an even one; some stakeholders are more powerful than others. For us, this means finding ways in which, for instance, immigrant based organizations, or, say, an organization working for teenage rappers will be treated fairly by representatives of civil servants.

Have You Discovered Any Novel Ethical Challenges or Responsibilities from Putting Work Online—Including Objects Created by People, Living and Deceased, from Cultures That Are Different than Your Own?

Anne Balsamo

When creating applications based on a sensitive topic, such as HIV-AIDS, it is not surprising that special care is needed when deciding how to make the information available for novel contributions from "the crowd." We refined our approach to crowd-sourcing archival information, to consider how to architect a "community-sourcing" approach instead. This shift, from the crowd to the community, represented a shift in our thinking about the nature of the "publics" that we wanted to engage in the AIDS Quilt Touch archival effort. This shift required us to think differently about what counts as "openness" in online applications.

Elizabeth Losh

Yes, and in some ways my answer might seem to contradict my response to the first question in this chapter. I often think that getting republication releases for digital artifacts can lead to some productive conversations, even if they are often legalistic. For example, in my first book *Virtualpolitik*, I wanted to show a screen shot from a PowerPoint slide show created by a military officer who was killed in Iraq. Speaking to his father was certainly awkward and painful, but it is important to understand that digital content-creators are complex human beings who might be speaking to multiple audiences and pursuing multiple purposes. When things I've written in blogs have shown up in news stories, as though the person actually interviewed me, I do think reporters are appropriating my work unethically, so I wouldn't want to similarly ignore the autonomy and ethos of others.

As to questions of cultural difference, that is a huge issue, particularly if you are interested in thinking about tactics in digital activism comparatively. When I was writing about the informational labor of feminists in India struggling with metadata naming conventions at the basic level of agreeing to use common hashtags in protest campaigns after the Delhi Rape Case, I tried to make sure that everyone I interviewed could read what I had written and see how they were being presented. Often I did this by inviting them to vet material in articles first published in blog posts. Asking someone to read an academic article can be an imposition but asking someone to read a blog post might be less demanding, and people who run NGOs or arts organizations are more likely to be appreciative of the signal boost for their work.

Because FemTechNet is an international collective, one of the things we can do is try to reverse flows about who gets to speak and on what terms. This is something that T.L. Cowan has been amazing about, making sure that workshop participants in Bangalore don't have to be up in the middle of the night to take part in conversations. Sometimes people at U.S. universities need to be the ones to stay up.

Heidi Rae Cooley

Perhaps the most important ethical responsibility we face is ensuring that students are prepared for and feel confident about speaking conscientiously and in very concrete terms about the nature of the projects they're helping to develop and the kinds of affects those projects aim to produce beyond the classroom. Because our class enrolls students from both computer science and the humanities, we find ourselves attending to diverse intellectual strengths and interests. How do we facilitate conversations about, for example, race and socio-economics, in ways that do not alienate students but rather empower them to consider the very material ways in which race and socio-economics have shaped the local landscape—and continue to do so? How do we guide them as they translate this understanding into content and mobile app functionality for the touchscreen device? In the case of the Ward One application, this is complicated by the fact that we're asking students to interact with individuals who witnessed the historical events the mobile app presents. Our interviewees are in their sixties and seventies, so we are also thinking about how to be attentive to generational differences, especially since many of them are less familiar with the technology and the logic of mobile applications—translation is key to gaining their trust. In sum, for us, "putting the work online" mostly raises the stakes on ethical considerations familiar to several kinds of classrooms.

Eric Hoyt

If someone else invests the time and money to digitize a public domain text, is it ethical to quietly copy the file and re-host it on your own site? David Pierce and I decided the answer was no. As a result, the Media History Digital Library only includes works that we helped to digitize, works in which the digital collection gave us permission (such as the Prelinger Library's lengthy run of *Business Screen*), or works that are part of our same open, shared hosting environment at the Internet Archive. Additionally, in all of these cases, we provide attribution to the original collections and sponsors.

The ethical questions relating to unpublished works can be even more challenging. To publish something, by definition, requires making it available to some public. But what about works that were not intended to be openly public? If researchers have to travel to a particular archive to view a collection of handwritten letters, then putting those letters online would certainly improve access. But is it ethically appropriate to make these private artifacts immediately accessible to the whole world? We might say yes in the case of a head of state, but no in the case of a less public figure. As the MHDL is presented with new collections to digitize, these are some of the questions we will continue to work through.

Tara McPherson

The case I noted above is certainly relevant here. We have also worked on projects that address this very issue. Anthropologist Kim Christen worked with others to create *Digital Dynamics Across Cultures* for *Vectors*. This piece powerfully explores the very western mindset that underpins techno-utopian notions like "information wants to be free." *Digital Dynamics* models non-western knowledge practices from indigenous Australia. In the Warumungu aboriginal community in Australia, certain images or video can only be viewed by those who have particular ties to those materials. As the user navigates Christen's piece, he may occasionally be barred from accessing a photograph or video. For indigenous peoples whose cultural production has often been raided for colonial and capital gain, openness and access are not inherently good. While openness can certainly be a good thing (as is often the case with open-access publishing or open software), it also has strong ideological underpinnings that we would do well to attend to with care. Projects like *Digital Dynamics* remind us of those ethical commitments.

Have You Ever Been Unexpectedly Drawn into an Applied Media Studies Project as a Participant in Ways That Redefined Your Understanding of Your Role as Scholar, or Expert, or Lead Investigator? How Did You Respond and Has This Changed the Way You Work Now?

Anne Balsamo

My identity is constantly shifting and morphing as the project itself unfolds. We are entangled in interesting ways at every moment. In my role as project manager and grant writer, I simultaneously "author" the project as I write documents and agendas and also as I interact with team members, grant representatives, or phantasms of intended users. The project does not precede me; I enact it, as do the others who participate. In this sense, the project is a boundary object; our identities are the enabling foundation for the project, and are also consistently reconfigured through it.

Elizabeth Losh

I think this happens all the time, particularly when you are crossing disciplines, transgressing boundaries between universities and their surrounding communities, or challenging the techno-missionary narrative in which the academy can be the only site of knowledge production. Because a lot of this happens in the context of developing bonds of friendship and hospitality, there is also an affective component that I sometimes feel I don't perform as well as I would like to. Luckily, I am fortunate to work with amazing activists like Jasmeen Patheja and Nishant Shah, who also happen to be creative artists and media scholars, and we can keep up with each other's lives and the banal and quotidian aspects of life through Facebook. Sometimes it's just channel checking, and sometimes it is affirmation, but a lot of my collaborators I think of as genuine friends rather than just Facebook friends.

Patrick Vonderau

Every single one of the projects I have initiated or participated in has redefined my understanding of my role as a scholar. That is exactly why I find them so interesting. Your question hints at the challenging aspects of such a redefinition but since my own career started at the margins of mainstream media studies— in an art school context—I have come to love working at the fringes.

Heidi Rae Cooley

Both *Ghosts of the Horseshoe* and the Ward One application—projects I would not have imagined before arriving at the University of South Carolina—serve as points of departure for my current theoretical thinking about habit-change and the relevance of American pragmatist Charles Sanders Peirce to the shape of that thinking. My work on these projects has also positioned me at Carolina as someone to consult regarding software development for mobile touchscreen devices. For example, I have recently been asked to take over as the project lead for *Handheld Art*, a software application for mobile touchscreen devices to be used in secondary classrooms for the purposes of encouraging "hands-on" exploration and study of artworks. And colleague and poet-laureate for the city of Columbia, Dr. Ed Madden, has approached me about meeting to discuss a mobile poetry application. Unexpectedly, then, my applied work has broadened my research horizons and newly credentialed me as a project manager.

Tara McPherson

Collaborating with the core *Vectors* team for over a dozen years has dramatically altered the way that I work. Every project we develop underscores for me the value of bringing together folks with very different skillsets and talents. While I still write single-authored scholarship, I find myself actively seeking out projects that are team-based and that reach beyond the university. I've also brought these collaborative practices into my classrooms and now often teach courses that engage students with community partners and in team-based learning. A recent example of this is a graduate seminar on "Activism in the Digital Age." In it, students worked in small teams to select and design our case studies and to implement actions outside of the classroom.

Reference

Willinsky, John. (2005) *The Access Principle: The Case for Open Access to Research and Scholarship*, Cambridge, MA: MIT Press.

13

TECHNOLOGY AND LANGUAGE, OR, HOW TO APPLY MEDIA INDUSTRIES RESEARCH?

Patrick Vonderau

What does "applied media studies" actually mean? Sometimes it simply means to speak of a study program that transfers academic knowledge to practical or functional contexts. For instance, the Dublin Institute of Technology offers an "MA in Applied Media Studies" that teaches "academic theories and concepts to help learners develop creative solutions to communications problems." "Applying" media studies here means matching academic expert knowledge to real-world situations, as opposed to the abstract or purely theoretical work allegedly done at, say, traditional humanities departments. One also speaks of applying a theoretical framework to an empirical case study, as in communications research that uses social theory to analyze the work of television journalists in order to make normative statements about news production (Perrin 2011). Proponents of "Critically Applied Media Studies" underline the role that scholarship can play in monitoring the political economy of media, for example, devising approaches that translate "media studies into social action." Such approaches are to be used "in the world outside of the classroom" for purposes of "education, resistance, and reform" (Karnik 1998; Ott and Mack 2010: 294–6). A related, if somewhat less politicized suggestion is to "apply traditional research methods" to digital media as a new object of study in order to identify "what new contexts have emerged to reinvigorate the central questions of our scholarship" (Sandvig and Hargittai 2015: 6).

What all these various ways of relating to applied media studies seem to have in common is that they are premised on the idea that media studies, in their traditional form and routine operations, are not sufficiently applied or applicable to society and the way society organizes through media. Identifying practical, societal, political, or methodological needs apparently not met by previous scholarship, these approaches aim to re-connect institutionalized knowledge production to a world caught, if not in disarray, at least in a process of accelerated transformation. Although this ethos and mission statement resonates well with the established communication studies agenda, especially as this agenda has distanced itself in its approaches from those of the humanities, it is equally reflected in the digital humanities (DH). Designating work done at the intersection of digital media and traditional humanistic study, the mission of DH has often been declared to bring both the tools and techniques of digital media to bear on traditional humanistic questions, and humanistic modes of inquiry to bear on digital media. Consequently, DH's programmatic visions of techno-science have promoted new forms of knowledge production going beyond traditional notions of academic authorship (net community vs. author), research practice (quantitative correlation of Big Data vs. qualitative analysis of small data), epistemic virtues (presentation/linking vs. argumentation/narration), or mode of publication (liquid net publishing vs. finished monograph or research article) (Hagner and Hirschi 2013).

It is debatable, of course, whether there ever existed such a divide between the humanities and the real world, between the inside and outside of the classroom, or between traditional and new objects of study. For those of us engaging in applied work of various kinds, there is neither reason nor need to recast the role of the scholar as that of a romantic hero or, even worse, as a service provider "in the knowledge and human-resources industry" (Collini 2012: 132). In an era of massified higher education, oppositions of theory/information or theory/know-how are frequently just used for re-modelling universities according to the needs of economic growth (Branston 2000; Gill 2014). Many of us also simply *have* to apply research findings, or claim to have done so, in order to match 'impact indicators' defined by university administrators and funding bodies. But applied or practical media work is not the opposite of, or an alternative to theory. In addition, few critical media scholars would go as far in applying their knowledge to real-world situations as historian Sergei Mironenko, the long-time director of Russia's State Archive, who recently lost his job for speaking out publicly against the misrepresentation and nationalist ideology in Russian blockbuster *Panfilov's 28* (Walker 2016). For most of us, interventionist approaches are desirable only as long as they do not entail serious consequences. At the same time, however, and despite these considerations, recent years have also clearly shown the need for what is less a substitution of practical applicability for theory, than what we may understand as a translation project between the two.

I like to think of digital media 'theory' and 'practice' as unlike cultures that interact on many levels. There is an extraordinary diversity of cultures that participate in digital media as a field of contemporary knowledge production. These different groups have distinct identities and differ in many ways, such as in their forms of work, analytic instruments, or modes of demonstration. Yet, there is still a complex dynamic by which common cause is made between and among them. Peter Galison, borrowing from anthropologist Nicholas Thomas, invokes the metaphor of trade in his history of microphysics, *Image and Logic* (1997), to describe how various subcultures within the physics community co-ordinated with one another without homogenization. "Two groups can agree on rules of exchange even if they ascribe utterly different significance to the objects being exchanged," Galison argues, "they may even disagree on the meaning of the exchange process itself. Nonetheless, the trading partners can hammer out a local coordination despite vast global differences" (783). Galison's famous account of the "trading zones" within modern physics provides a useful model for describing how various traditions of creative or practical work on the one hand, and academic work on the other can coordinate with one another without turning into one consistent mode of knowledge production. Thinking of applied media work as a trading zone rather than an emerging field or even discipline also redirects attention to the particular sites of exchange, and towards the "contact languages" or systems of discourse that develop within a framework of exchange locally.

Applied media work is a vast border or contact zone where exchange happens between various groups, including programmers, archivists, librarians, academic researchers, film-makers, or political activists. The 'zone' thus extends from interdisciplinary and intra-group (intra-academic) exchange to various forms of inter-group contact. This is important because of the often unintended dynamics or consequences that may evolve while crossing two boundaries simultaneously: the boundary in between disciplines on the one hand, and the boundary between academia and the world it observes on the other. One key site for experiencing these kinds of tensions, but also the sort of benefit applied media work may produce, is media industries research. Describing its mission as a "transdisciplinary conversation about the converging global media landscape," proponents of this new field have argued for a "more integrative approach" that would allow for bottom-up study of cultural production, foregrounding the role of individual agents within larger media structures (Holt and Perren 2009). This means importing participant observation, ethnography, shadowing, interviewing, and other micro-sociological or social science methods into humanities-based research traditionally premised on analyzing and interpreting texts (Caldwell 2008). It also means to engage in a direct conversation with those making texts, such as producers or distributors, marketers or advertisers, and all the new intermediaries that make up what is called the digital economy.

One of the recurring and widely acknowledged problems of this type of research is that its qualitative social research methods are of limited use in obtaining access to industries. "Access" here is not confined to the issue of making first contact or building rapport with an industry insider, which often turns out to be easier than expected. Rather, I understand access in the sense first proposed by anthropologist Laura Nader, as a form of "studying up in our own society," following a "normative impulse" to ask questions about the "mechanisms whereby far away corporations and large scale industries are directing the everyday aspects of our lives" (1970: 5). Nader's famous paper, written in the late 1960s, reflects a research ethos and an understanding of the democratic right to public access that have been largely lost in today's media studies, and especially so in works on digital media or the Internet. Although there is certainly no shortage of normative views on media, the impulse to publicly break down the barriers blocking access to corporate cultures themselves is now easily dismissed as naïve or off the point. Scholars aiming to publish the results of empirical research on media industries usually begin by signing non-disclosure agreements, anonymizing sources, and receiving litigation counseling. While research always needs to fully protect and respect its sources, especially in contexts as sensitive and confidential as this one, "studying up" the media industries invokes a power imbalance that has little to do with getting approval from your university's Research Ethics Committee. Applied media work, in my view, may contribute to exposing and partially eliminating this imbalance by devising different approaches and methods.

To underline the need for applied media work in this specific sense, as a critical 'turn to method' (rather than a complete re-tooling) within media industries research, we may contemplate the degree to which scholars in this evolving field are enticed to adopt the agenda, guiding metaphors, and promotional jargon provided by the industries themselves. Research on digital media and the Internet today is dominated by a small set of issues and terms understood to be more or less self-explanatory, such as "participatory culture," "big data," "platform responsibility," or "sharing economy." Revisiting these and other terms now, after more than a decade of "Web 2.0" scholarship, and in light of the last U.S. presidential election and "post-truth politics," does not exactly support the idea that they have helped academics much to explain how the "information society" would develop. To the contrary, it might be a useful starting point for a project to take any of these imperatives and ask "what would be done any differently in contemporary computational and networked media were that maxim reversed" (Fuller and Goffey 2012). In other words, rather than focusing on platforms, algorithms, participation, streaming, or sharing, for instance, we may want to ask who established such focal points in the first place, and what do they *not* speak about.

In part, the closeness of our analytical language to industry jargon can be seen as reflecting a tacit assumption among industry researchers that objectivity has to be reached through immersion, in the sense once proposed by social anthropologist Bronisław Malinowski (1922: 25, original emphasis), by "grasping the native's point of view, *his* relation to life, and to realize *his* vision of his world." There is, however, also considerable evidence that media companies strategically manage their relationship with academia. Like any other major industry, media industries seek to assuage concerns by co-opting the discourse of their critics: they promote themselves as responsible, enhance their reputation through forging strategic partnerships, and occasionally spread uncertainty and doubt (Kirsch

2014: 3). It is hardly coincidental that key impulses for setting the scholarly agenda on digital media and the Internet have come from scholars paid by Microsoft (e.g., Nancy Baym, danah boyd, Tarleton Gillespie), Intel (e.g., Genevieve Bell, Melissa Gregg), or by universities that depend on industry partnerships to a considerable degree (e.g., Henry Jenkins, Jean Burgess). Although this does not devalue individual research accomplishments, it raises the question why "applied research" cannot or should not be imagined outside such symbiotic relationships between industries and their observers, and about the value and ethics of media industries research as such. Sociologist Niklas Luhmann (1994) once argued that all forms of science, including use-oriented research, are operating as "distinction-generating programs" (Luhmann 1994: 645), and thus crucially depend on establishing a difference between their observations and their objects. Where there is no such difference we cannot speak of science, according to Luhmann.

To put it differently, and by coming back to the notion of the trading zone, exchange between the worlds of science and media requires more than borrowing words from the natives, or, in Galison's (1997) phrasing, "foreigner talk" (770). It requires a fully developed trading language (what linguists call creole) that not only functions in a more varied way but even includes "metalinguistic reflections, irony, humor, and the uncounted conventions that go with daily life" (ibid.). "Applied" media research, I argue, begins on the level of language. Technology is not the decisive factor; language use is. While Galison used the comparison with trading languages to retrospectively explain how subcultures within physics coordinated despite differences, we may also use the analogy prescriptively or at least in the sense of a thought experiment, as a suggestion for future work. For studying digital media industries within the humanities, our analytical language neither needs to import its central conceptual metaphors from the field nor to build its epistemic frameworks on computational techniques (cf. Verhoeven 2016: 166). Instead, we may begin re-building (or adjusting) such frameworks by thinking more critically about the relations between language, power, and agency, by engaging in more reflexive language use, by using language in conceptually productive or even subversive ways—poststructuralist and feminist approaches had a lesson there.

Following these impulses, the "normative" impulse to research the "mechanisms" of large scale industries on the one hand, and the skeptical impulse to build such research on language games and experimental epistemologies rather than, say, a belief in the powers of computing or technology, we came to develop, and then conduct, a collaborative project entitled *Streaming Heritage: Following Files in Digital Music Distribution* (2014–2018). This project brought together five scholars from four disciplines: a social anthropologist (Maria Eriksson), an ethnologist (Anna Johansson), an economic historian and Internet activist (Rasmus Fleischer), and two media scholars (Pelle Snickars and myself). What brought us together was an interest in using Spotify, the then globally hyped Swedish music streaming service, as a lens to explore some less obvious social, technical, and economic processes associated with digital distribution. Meeting over a lunch break in early 2013, our first intuition indeed was to *use* Spotify analytically, rather than transferring authority to the platform itself, as recent platform-oriented approaches do when they closely adhere to how social media companies like Twitter present themselves (Baym and Burgess 2016). De-centering the object of study, and strategically ignoring the cultural and economic power of a mighty industry that scholars are expected to "study up" to, we aimed at using Spotify as a simple means for accessing infrastructures whose materialities and procedures are not frequently talked about. This would require mixed and digital methods, an anarchist ethos of interventionism, and therefore, building on transdisciplinary exchange and potential cross-border conflict.

Several issues sparked our interest in Spotify. The service counts as the "biggest streaming service in the world" (Murgia 2016) and as Europe's most valuable "unicorn" start-up, worthy of exceptional political support and protection. Initially founded in the intent to mediate between Sweden's lively (but unauthorized) file-sharing scene and the commercial music industry, and using peer-to-peer technology, Spotify has pushed the idea of streaming as a solution to problems of the cultural and digital economies far beyond its own platform. Spotify has even been adopted by public service institutions such as Swedish Radio or The National Library of Sweden to define a socially viable, open access solution for cultural heritage. At the same time, however, the company also has gained considerable notoriety for introducing a revenue standard that puts independent musicians at a disadvantage, and for gradually transforming its music distribution network into a platform for collecting behavioral user data. Spotify never generated substantial revenue, but the company did receive eight rounds of equity funding and twenty-eight mostly American investors have turned the company into an asset traded on financial markets. As of today, Spotify operates like a traditional American media company, while retaining the benefits of financial and regulatory loopholes granted to European tech firms. In short, rather than epitomizing a solution to, say, the observed devaluing of cultural content in the digital economy, Spotify epitomizes a wide-spread "solutionism" (Morozov 2013) that raises questions about agency, power, and access to cultural heritage.

In our initial discussions, we quickly agreed that none of the common approaches to digital media would be particularly helpful in drawing the more holistic and critical picture of digital distribution we aimed at. Our project was neither meant to become another user study, nor did we want to confine our empirical materials to industry interviews, economic/industrial analysis, or field observation. In order to include in our investigation issues related to Spotify's historical change, internal contradictions, hidden agendas, failures, and "multiple ontologies" (Zuiderent-Jerak 2015: 159), we decided to go beyond its platform, speculations about the power of algorithms, or contemplations about "on-demand culture" (Gillespie 2010, 2014; Tryon 2012). What we thought of was something more akin to the tradition of breaching experiments in ethnomethodology: a concerted effort to 'break' into the hidden infra-structures of digital music distribution in order to study its underlying norms and structures (Garfinkel 1967). Taking our cue from economic and social anthropology with its interest in the "social life" of things (Appadurai 1986), and inspired both by the infrastructural turn in media studies and recent writings in the field of Science and Technology Studies, we developed an interventionist approach that would follow the transformation of audio files into streamed experiences in the simple way a postman would follow the route of a parcel—all the way from packaging to delivery. This approach be-came an intervention by interfering with information technology (IT) logic that "following files," of course, is a technical impossibility. Our project thus from the very beginning took a stance against this common position that sets the values, ideologies, and concepts of software engineers, IT specialists, or start-up entrepreneurs as a default, letting them define what digital culture is, and how it should be analyzed.

The Swedish Research Council encouraged us to develop this "gonzo-strategy" (as one reviewer put it) and in early 2014 granted roughly one million USD in research funding. In making "following files" the guiding metaphor for our project, we consequently came to engage in forms of exchange that would differ markedly from the ways mainstream communication or media studies projects would 'deal' with digital industries. The experimental setting for studying distribution processes included the distribution of self-produced music/sounds through Spotify; the documentation and tracing of Spotify's history through constantly changing interfaces; the charting of the industries, and interviews/interventions with providers at the backend of such services; and interviews with Spotify management that were auto-ethnographically self-observed in order to expose the researcher's problems in studying up. During this research process, we asked: How is streamed music commodified? What sounds are perceived as music (or not) according to Spotify and adjacent aggregating services? How are metadata generated, ordered, and valued—and what kind of metadata are actually available? What normative world views are promoted and materialized by streaming architectures? How is the social con-tract between listener communities, artists, and collecting societies changing? What kind of infrastructures proliferate behind the surfaces of on-demand services? In what ways does music as data drive botification where automated scripts feedback and reshape listening behaviors? (Figure 13.1)

One of the experiments we started with was the setting up of several ficti-tious record labels that would allow us to observe how aggregators and other intermediaries define sounds as music. The objective here was to use meth-ods that not only investigate the social world, but engage it; that not only report on events, but act on them; that not only happen, but make things happen; and that are not only the source of making, but also unmaking,

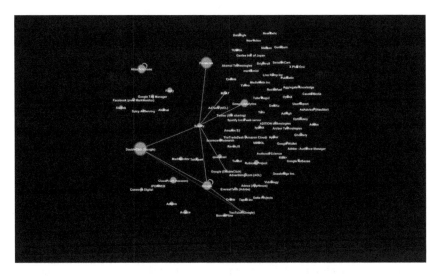

FIGURE 13.1   An experiment using software tools to "sniff out" the communication between Spotify and its ad-supplying partners.

Source: Image by Patrick Vonderau.

remaking, and picking apart (Lury and Wakeford 2012: 6–9; Eriksson 2016). Team member Maria Eriksson, for instance, released an EP on Spotify with sampled breakfast noise (called *Frukost*), while Rasmus Fleischer published *Election Music*, an album consisting of one single electronic tune that was automatically modified according to the voting results in the 2014 Swedish government election. The aim with these releases of course was not to create a hit; rather, we departed from the notion that fraud is as inherent to digital technologies as underground economies are inherent to their business models (Lobato and Thomas 2015). Spotify is full of spam, fakes, and clones; in releasing 'polluted' sound materials we aimed to test the limits of content management systems of streaming platforms, and aggregators such as CDBaby, TuneCore, or Record Union in particular. As Eriksson (2016) observed, music aggregators operate on a logic of abundance that would accept kitchen noises as music provided that an aggregation fee is paid. But interaction with aggregators also provided us with tools offered to musicians and labels that both allowed us to monitor how our sounds perform, and to inspire other kinds of experiments.

These include, among others, "Turing tests" of Spotify's monetization system (Snickars and Mähler 2016). Creating a bot, or small software script, that would listen to Abba's *Dancing Queen* on a fixed repetition schedule running from ten to a couple of thousand times on repeat, the experiment once more showed how easy it is to game the digital value chain. We also used large numbers of bots to listen to our self-produced sounds, thus short-circuiting the system, as it were. Other bots were designed along gendered norms of music listening and used to expose the gendering of culture in the streaming world (Eriksson and Johansson 2015). In a second stage of the project, we even began using the results of these and other experiments in order to stir up a public debate about Spotify's role and function for cultural heritage. For instance, we invented SongBlocker, a tool that gives you "100% ads when listening to Spotify. No music. No disturbance," accompanied by the release of SongBlocker Pro, the premium version, enabling paying users to listen to ads offline, receive targeted recommendations of the hottest ads, and listen to the ads their friends are listening to. The ironic and public play with Spotify's attempts to increase the platform's value by introducing micro-targeting or "programmatic advertising" here went along with a detailed study of Spotify's ad networks and ad tech intermediaries more generally (Vonderau 2017).

While our research will have limited consequences when it comes to changing normative assumptions about digital media or their underlying economic or political rationale, changing the language through which we interpret such assumptions or plans is useful in itself. It helps us regain scholarly authority in a time when language is bent to the corporate will, used to disinform and to intimidate. Far from being impartial onlookers, digital media companies such as Facebook, Twitter, Google, or Spotify play a crucial part in redesigning social worlds in a way that is neither transparent nor foreseeable. Who would have predicted that psycho-graphic models used to further the impact of the Trump campaign were inspired by the behaviorist user targeting of a music streaming service like Spotify (Greenberg et al. 2016)? What media studies cannot afford, at this polycultural moment of its history, is to treat engineering and other cultures as something foreign to its own agenda and interests. Applied media studies, in this context, means to engage in creating contact languages or systems of discourse that enable us to organize knowledge in critically productive ways. As Gayatri Chakravorty Spivak (2008) put it,

> the production of language is our practice. The received dogma asks that our language be pleasant and easy, that it slip effortlessly into things as they are. Our point of view is that it should be careful, and not take the current dogmatic standard of pleasure and ease as natural norms.
>
> *(382)*

References

Appadurai, Arjun, ed. (1986) *The Social Life of Things. Commodities in Cultural Perspective*, Cambridge and New York: Cambridge University Press.

Baym, Nancy, and Jean Burgess. (2016) "A Biography of Twitter," Paper Delivered at the *Association of Internet Researchers Conference*, 5–8 October 2016, Berlin, Germany.

Branston, Gill. (2000) "Why Theory?," in Christine Gledhill and Linda Williams (eds.), *Reinventing Film Studies*, London: Arnold, 18–33.

Caldwell, John Thornton. (2008) *Production Culture: Industrial Reflexivity and Critical Practice in Film and Television*, Durham, NC and London: Duke University Press.

Collini, Stefan. (2012) *What Are Universities For?* London: Penguin.

Eriksson, Maria. (2016) "Notes from A Scientific Record Label," Paper Delivered at the *Society for Cinema and Media Studies Conference*, 30 March–3 April 2016, Atlanta, GA.

Eriksson, Maria, and Anna Johansson. (2015) "Keep Smiling!: Time, Functionality, and Intimacy in Spotify's Featured Playlists," *Cultural Analysis* (forthcoming).

Fuller, Matthew, and Andrew Goffey. (2012) *Evil Media*, Cambridge, MA: MIT Press.

Galison, Peter. (1997) *Image and Logic: A Material Culture of Microphysics*, Chicago, IL: University of Chicago Press.

Garfinkel, Howard. (1967) *Studies in Ethnomethodology*, Englewood Cliffs, NJ: Prentice Hall.

Gill, Rosalind. (2014) "Academics, Cultural Workers and Critical Labour Studies," *Journal of Cultural Economy* 7(1): 12–30.

Gillespie, Tarleton. (2010) "The Politics of Platforms," *New Media & Society* 12(3): 347–64.

———. (2014) "The Relevance of Algorithms," in Tarleton Gillespie, Pablo J. Boczkowski, and Kirsten Foot (eds.), *Media Technologies: Essays on Communication, Materiality, and Society*, Cambridge, MA: MIT Press, 167–194.

Greenberg, David M., Michal Kosinski, David J. Stillwell, Brian L. Monteiro, Daniel J. Levitin, and Peter J. Rentfrow. (2016) "The Song Is You: Preferences for Musical Attribute Dimensions Reflect Personality," *Social Psychological and Personality Science* 7(6): 597–605.

Hagner, Michael, and Caspar Hirschi. (2013) "Editorial," in Andreas B. Kilcher, Caspar Hirschi, David Gugerli, Jakob Tanner, Michael Hagner, Patricia Purtschert, and Philipp Sarasin (eds.), *Nach Feierabend. Zürcher Jahrbuch für Wissensgeschichte 9*, Zürich: Diaphanes, 7–11.

Holt, Jennifer, and Alisa Perren, eds. (2009) *Media Industries: History, Theory, and Method*, Chichester: Wiley-Blackwell.

Karnik, Niranjan S. (1998) "Whose News? Control of the Media in Africa," *Review of African Political Economy* 25(78): 611–23.

Kirsch, Stuart. (2014) *Mining Capitalism: The Relationship between Corporations and Their Critics*, Berkeley: University of California Press.

Lobato, Ramon, and Julian Thomas. (2015) *The Informal Media Economy*, Cambridge: Polity Press.

Luhmann, Niklas. (1994) *Die Wissenschaft der Gesellschaft*, Frankfurt am Main: Suhrkamp.

Lury, Celia, and Nina Wakeford, eds. (2012) *Inventive Methods: The Happening of the Social*, New York and London: Routledge.

Malinowski, Bronisław. (1922) *Argonauts of the Western Pacific: An Account of Native Enterprise and Adventure in the Archipelagos of Melanesian New Guinea*, London: Routledge & Kegan Paul.

Morozov, Evgeny. (2013) *To Save Everything, Click Here*, London: Allen Lane.

Murgia, Madhumita. (2016) "Spotify Crosses 100m Users," *The Telegraph*, UK.

Nader, Laura. (1970) "Up the Anthropologist: Perspectives Gained from Studying Up," in Dell H. Hymes (ed.), *Reinventing Anthropology*, New York: Random House, 284–311.

Ott, Brian L., and Robert L. Mack. (2010) *Critical Media Studies: An Introduction*, Chichester: Wiley-Blackwell.

Perrin, Daniel. (2011) "'There Are Two Different Stories to Tell': Collaborative Text-Picture Production Strategies of TV Journalists," *Journal of Pragmatics* 43(7): 1865–75.

Sandvig, Christian, and Eszter Hargittai, eds. (2015) *Digital Research Confidential: The Secrets of Studying Behavior Online*, Cambridge, MA: The MIT Press.

Snickars, Pelle, and Roger Mähler. (2016) "SpotiBot: Turing Testing Spotify," Paper Presented at the *Digital Humanities Conference*, 11–16 July 2016, Kråkow, Poland.

Spivak, Gayatri Chakravorty. (2008) *In Other Worlds: Essays in Cultural Politics*, New York and London: Routledge.

Verhoeven, Deb. (2016) "Show Me the History! Big Data Goes to the Movies," in Charles R. Acland and Eric Hoyt (eds.), *The Arclight Guidebook to Media History and Digital Humanities*, Sussex: Reframe Books, 165–83.

Vonderau, Patrick. (2017) "Scaling Up: Streaming, Venture Capital, and Programmatic Advertising," in Denise Mann (ed.), *Content Wars*, New Brunswick, NJ: Rutgers University Press (forthcoming).

Walker, Shaun. (2016) "Russian War Film Set to Open Amid Controversy," *The Guardian*, UK, 23 November.

Zuiderent-Jerak, Teun. (2015) *Situated Intervention: Sociological Experiments in Health Care*, Cambridge and London: The MIT Press.

14

TRANSFORMING THE URBAN ENVIRONMENT

Media Interventions, Accountability and Agonism

Bo Reimer

The New Media Research Landscape

The media landscape has changed tremendously. New companies—Apple, Google, Facebook, and so on—have entered the landscape, reshaping it from the ground, and disrupting the business models of the traditional media companies. Media as such are increasingly being designed, produced and consumed collaboratively. And as evidenced in the Internet of Things discourse, more and more objects become media.

These changes put new demands on media research. How can we as media researchers best make sense of this new situation? What can a media researcher actually do in this changing media landscape? In order to give an answer to that question, it may be reasonable to start by reflecting upon what media research is today. It is, of course, presumptuous to speak of media research as a singularity; there are different traditions and approaches. But unquestionably most work belongs within the humanities and the social sciences, and is analytical/ critical. As a media researcher, you theorize, you analyze, you interpret, and you produce conclusions, stating how things "are." Additionally, you may add normative statements based on the analysis, stating for instance that the situation is not optimal ("the working conditions for female journalists are terrible," "the quality of the media output could be higher," etc.). You may also add creative, constructive conclusions, stating, "This is how it should look, and if you do this…"

The latter stance makes explicit an interest in not only stating how things are, but also in considering the possibilities for change. In that sense, it is a way of taking a stance; rather than just "objectively" observing and analyzing, you take part. But it is a way of doing it from a distance. It is also based on the notion of only reacting to things already happening.

Taking such a stance is valuable. But there are problems with it. Is it really possible to stay on the outside as an objective onlooker, not shaping the experiences? And even if it would be possible: is it really the best position to take? That is not the view put forward here. In today's media landscape—a landscape I would like to denote as a collaborative media landscape (Löwgren and Reimer 2013)—it is easier than ever before to take both a collaborative and an interventionist stance. As I will argue later, it is also more important.

Why is it easier? In the same way that cheap, easy-to-use technologies are making it easier both for traditional media producers and for what used to be called media consumers to collaborate, it has also become easier for media researchers. The tools that are available to producers and consumers are also available to us.

And media researchers are no longer—if they ever were—the only academics interested in media. Today researchers dealing with the media also come from media technology, interaction design, informatics, human-computer interaction, and computer science. Increasingly, there is also arts-based research of highest relevance. Thus, the academic field of media research has broadened immensely, and the possibilities for conducting collaborative research have consequently grown.

Additionally, there are actors outside of academia to be brought into the knowledge processes. These are stakeholders with sometimes similar, sometimes different interests, and these are stakeholders with other kinds of competencies. What is important to note, however, is that in creating these kinds of collaborations, it makes less and less sense talking about "us" and "them." As Bruno Latour (2004) argues:

> The sharp distinction between, on the one hand, scientific laboratories experimenting on theories and phenomena *inside* their walls, and, on the other, a political *outside* where non-experts were getting by with human values, opinions and passions, is simply evaporating under our eyes. We are now all embarked in the same collective experiments mixing humans and non-humans together—and no one is in charge.
>
> *(4)*

Therefore, what we are seeing are the possibilities for a research approach to collaborative media and to the contemporary media landscape in which media researchers from the social sciences and the humanities join forces with technology-oriented researchers, with practice-based and arts-based researchers, and with non-academic actors to conduct real-life experiments "in the wild."

What is crucial with this approach is the orientation toward intervention and change, where creative work is part of the knowledge production. And where the focus is as much on how something could be, as on how something is.

Collaborative Media Interventions at The School of Arts and Communication, and at Medea, Malmö University

Malmö University is a new university; it started in 1998. One of the ideas behind the university was to think in new ways in relation to the creation of multidisciplinary units. One of these units was The School of Arts and Communication. What characterizes the school is a focus on design, media, arts, and technology, on cross-disciplinary competence, and on the integration of theory with practice. Points of departure were taken from the Bauhaus school, as well as from the ideas of John Dewey (1916/1997, 1927/1991). Donald Schön's writings on the reflective practitioner also served as important inspiration (1983). In 2009, The School was complemented by Medea, an experimental lab dealing with collaborative media, design, and public engagement.

Taking part in the creation of The School of Arts and Communication, I started to work for Malmö University one year before its official inauguration. What attracted me to The School was the possibility of doing media studies work that extended beyond the traditional theoretically informed and empirically based work that dominated the field. This possibility, of course, grew out of the possibilities inherent in the creation of a new environment; an environment to which people came precisely because they wanted to do work that challenged what they had done previously. For the teachers and researchers coming from the humanities or the social sciences, this meant getting your hands dirty in the sense of making things—taking part in productions—or in the sense of intervening in ongoing everyday life processes. For the teachers and researchers coming from design or the arts, it meant placing their practice based work in new theoretical-critical contexts.

I will here give some examples of the work that has been carried out, describing briefly some projects involving, in different constellations, researchers, external actors and the general public. They have a focus on collaboration and on media, with the perspective on media being a broad one.

KLIV: Collaborative Media Design in a Medical Setting

Intensive care units at hospitals are places where mistakes may have lethal consequences. They are also units with complex, technical equipment, and it is crucial that the medical staff working at the units know how to handle them in emergency situations.

How do you describe how complicated equipment function? The traditional way is to create manuals. But as everybody knows who has tried to follow such manuals, that can be very tricky. It seems to be very difficult to write manuals in ways that make them useful; quite often it seems as if the manuals are written from another perspective than the one of the user. Writers of manuals obviously try to make the manuals useful in concrete instances. But much use is situation-specific, and without having a manual swelling into hundreds and hundreds of pages it is very difficult to cover all possible situations.

But there is an additional problem. Not only is the use situation-specific. It is also context-specific; it is dependent on how the particular equipment is related to other equipment in very specific settings. And it is dependent on the shape the actual equipment is in. Even though not faulty, due to heavy use equipment may need special treatment in order to function. Maybe one button needs to be pressed much harder than other buttons. Or on occasion, maybe some part of the equipment needs a swift kick simultaneously with the pushing of a button. These things a manual can never help you with.

So what is the solution? If the knowledge required is context-specific, then it should be people knowing the specific instruments that provide the knowledge, and it should be done in a context-specific way. That was the solution coming out of a collaboration between two of my colleagues at Malmö University—Erling Björgvinsson and Per-Anders Hillgren—and the medical staff of the intensive care unit at Malmö University Hospital. The project was called KLIV—an acronym in Swedish for "collaborative learning within healthcare".

The collaboration started out as part of the unit's work on competence development in relation to information and communication technologies; for this reason, they brought in Björgvinsson and Hillgren, two interaction designers experienced in facilitating participatory design processes. After a number of phases, including ethnographic studies of the environment, dialogues, and explorative experiments, what the researchers and the staff members together found was the potential inherent in creating location-based videos dealing with the staff members' everyday life situations, especially those involving the use of complicated equipment.

Initially, the researchers filmed the staff members using the equipment, but in the next step, also the filming was handed over to the member. This had a number of advantages. First, knowing each other, the recording situation was more comfortable. And second, since both were knowledgeable about the equipment, the staff member recording and the staff member acting could discuss in detail what was happening while the recording was made. Having been created, the videos could then be viewed on handheld computers (PDAs) when using the equipment.

The system became very successful. It was in use for many years and won a number of awards. It was also exported to other hospitals (Hillgren 2006; Björgvinsson 2007; Löwgren and Reimer 2013: 87–92).

See the Future: Developing the Urban Environment Together

In 2007, the traffic environment unit of Malmö city started a project dealing with getting inputs of citizens concerning how to reduce car traffic in the center of town. The project started by putting an open question on the city's official webpage. The question quickly got more than 1,000 responses. In the next step, the city started a collaboration with a small design and innovation studio called Unsworn Industries (a studio whose founders had connections to the research groups at The School of Arts and Communication). Together they started a campaign called "See the Future."

The idea behind the campaign was to move the citizen input from email suggestions to something more concrete. The design studio therefore created a device called the Parascope.

The device functions as a future periscope, through which you can view what the city or square in front of you may look like in the future. You can move the periscope around, you can zoom and access text annotations. And by pushing a button you can switch different visions of the place in question (Figure 14.1).

The visions were created by architects, artists, graphic designers, and cultural geographers, and the idea was to get citizens more concretely involved in what their city should look like. Rather than just getting the opportunity to comment on developments already decided upon, working this way meant getting citizens involved much earlier in a process. Furthermore, working with augmented reality overlays rather than with maps or models made it easier for

FIGURE 14.1 The Parascope.
Source: Image by Unsworn Industries, used with permission.

citizens to grasp what the proposals actually could lead to—and being able to view the alternatives out in the actual physical environment that was about to change also added to the realism.

The See the Future project was in many ways successful. It created a link between the city officials responsible for urban planning and some of the citizens of Malmö. However, from the perspective of citizens, this could still be considered a top-down process. True, citizens were given the opportunity to comment on different alternative visions for the future of their city, but it was still a question of having to choose between alternative visions, already created for them by professionals. The citizens themselves had no input in those visions.

In order to overcome this problem, in a second stage, and in collaboration with researchers at The School of Arts and Communication, the project therefore turned into a more participatory planning project. In an office in one particular Malmö neighborhood, where changes were about to happen to a square close to the office, a panorama studio was set up. In this studio workshops were organized with citizens who there could create their own versions of the future. These visions were then shown in a Parascope on the square. In further experiments, the studio was moved outside to the actual square, in order to make the connection even stronger.

The project is now concluded, but the Parascope has been used in more than 20 different settings in Sweden and in other countries (Löwgren and Reimer 2013: 59–67).

Malmö City Symphony: Collaborative Media Productions, and the Re-Mixing of the City

Between 2008 and 2010, I led a research project called "Designing for Collaborative Cross-Media Production and Consumption." It was financed by the Swedish Knowledge Foundation. The project contained a number of subprojects, one of which dealt with the creation of a collaborative media production based on the documentary genre of city symphonies. This genre, with its origins in films of the 1920s such as Dziga Vertov's "The Man with the Movie Camera" (1929) and Walter Ruttman's "Berlin: Symphony of a Metropolis" (1927) is in no way tied to the notion of collaboration. Quite the opposite; each of these 1920s films is highly associated with its powerful (male) auteur.

However, what interested the leaders of this particular subproject, Erling Björgvinsson and Richard Topgaard, was the location-specificity of these films. Most of the films focus on a particular city over the course of a given day, catching its ebbs and flows.[1] In other words, there is a fairly defined area to represent—or maybe to perform.

What happens if you remove the lone auteur and replace him with a number of film makers, amateurs and professionals, and then let them produce moving images of a particular city irrespective of each other; images that could be combined and used in many different ways and at different occasions?

This was the starting point of the Malmö City Symphony project. In a collaboration between Medea, the cultural institution Inkonst, and the documentary film club Doc Lounge, the word was spread about the possibility for anyone with access to a video camera or a smartphone to shoot moving images of Malmö and upload them on a webpage in order to participate in the production. No sound was needed. The clips were available on the webpage for six weeks, during which time it was possible for anyone who registered on the page to remix them. All in all, sixty clips were produced.

A month later, the material was used as the basis for a live event at Inkonst. The clips had been given to two VJs who during the event remixed and manipulated them in real time. Their work was complemented by two electronic musicians who added sound—some of which was pre-recorded, some improvised. If you were at Inkonst (about a hundred people were there) you could watch the VJs and the musicians performing in front of a giant screen showing the film as it was being performed. But you could also follow the event online: Richard Topgaard, one of the project leaders, did his remixing of what was happening live, and broadcast that over the net (Figures 14.2 through 14.5).

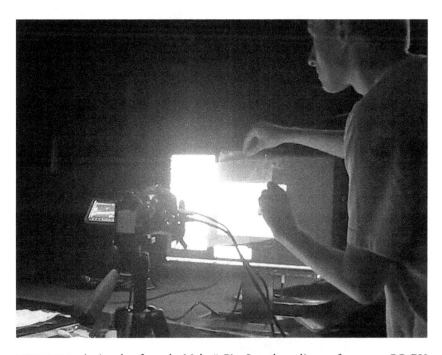

FIGURE 14.2 Action shot from the Malmö City Symphony live performance. CC–BY Malmö City Symphony. http://citysymphony2009.blogspot.se/.

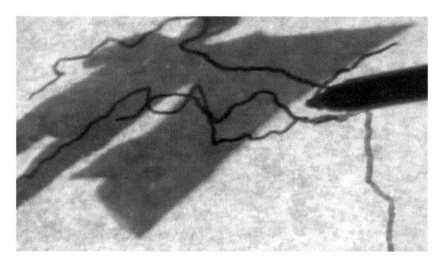

FIGURE 14.3 Action shot from the Malmö City Symphony live performance. CC-BY Malmö City Symphony. http://citysymphony2009.blogspot.se/.

FIGURE 14.4 Screenshot from the Malmö City Symphony live performance. CC-BY Malmö City Symphony. http://citysymphony2009.blogspot.se/.

The individual clips are currently not available online, but the recording of the live event can be found at http://citysymphony2009.blogspot.se/ (Löwgren and Reimer 2013: 124–30; Björgvinsson and Høg Hansen 2016).

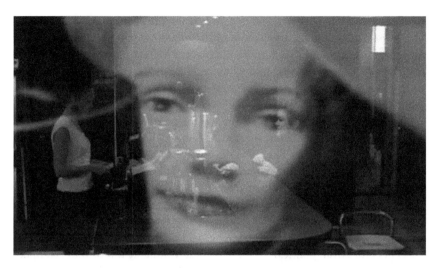

FIGURE 14.5   Screenshot from the Malmö City Symphony live performance. CC-BY
Malmö City Symphony. http://citysymphony2009.blogspot.se/.

The Politics and Ethics of Applied, Collaborative Media Research

In the previous section I described three projects that in different ways have been
tied to the research environment of The School of Arts and Communication and
the research lab Medea at Malmö University. There are other projects that I could
have chosen as well; these environments have been very productive during the
last fifteen years (Björgvinsson, Ehn, and Hillgren 2012; Löwgren and Reimer
2013; Ehn, Nilsson, and Topgaard 2014).[2] But these three give especially interesting insights into what it means to work collaboratively and with an intervention-oriented approach to media. They also all deal with the urban environment.

In the work we carry out, a common point of departure for many, albeit
not all, projects is the idea of participation. But not participation in general,
of course. It is about participation in socially, politically, or culturally relevant
projects, where we may join already ongoing projects or we may be the ones
starting processes that would not have happened otherwise. The Malmö City
Symphony project is an example of a project that originated with us, whereas
See the Future was a project that we joined in a relatively late (although very
important) phase.

From one perspective, the Malmö City Symphony project was primarily a
cultural project. It placed itself within the confines of a very clearly delineated,
high cultural film genre, with one aim being to develop that genre. That is an
obvious cultural objective. But what was especially interesting was the way
the project tried to develop the genre: not by doing just one more film with a

somewhat different perspective, but by remaking the genre into a collaborative film experiment, and by emphasizing fluidity and openness. The idea was to make possible an indefinite number of different Malmö City Symphonies, and make possible continuous remixes of the ones previously made. But in addition to this cultural objective, an important part of the project was the event at Inkonst, which was a social gathering as well as a cultural one.

The KLIV project may be regarded as an exemplary participatory design project. Different stakeholders meet and together they identify a problem and come up with a solution. However, "participation is risky," as the title of the edited volume by Liesbeth Huybrechts (2014) spells out. You have to negotiate with people that may be driven by other values or interests. Projects can fail. These are things you have to take into account. So you have to think through how to deal with them.

In these matters, the concept of agonism has been very useful for us. The concept comes from political theorist Chantal Mouffe. It was introduced in her book, "The Democratic Paradox" (2000), and has since been further developed in other writings (Mouffe 2013).

For political deliberation, Mouffe makes the distinction between adversaries and enemies. An adversary is an opponent you do not agree with, but with whom you share the democratic principles of freedom and equality. Thus, an adversary is a person you can discuss and communicate with. An enemy, on the other hand, is a person with whom there are no grounds for communication. Following upon this, she distinguishes between agonism and antagonism. A struggle between adversaries is an agonistic struggle. A struggle between enemies is an antagonistic struggle.

Two important points follow on this. First, a struggle with an enemy is meaningless. For it to become meaningful, the relationship has to turn from an antagonistic one to an agonistic one. This is not easy, but it is possible. Second, the "normal" political situation is one of agonism, not of consensus—and that is a good thing! Political actors normally do not agree. But the point is not to reach complete consensus. Societies develop through disagreement and debate. As long as we agree on the possibilities for disagreement, we are fine.

Mouffe's concept of agonism has been influential in social and political theory, but it has also been picked up by researchers within other fields, for instance by participatory design researchers (Björgvinsson et al. 2012; DiSalvo 2012). And that makes perfect sense, since the kinds of projects participatory design researchers conduct involve a number of stakeholders with different interests.

Of the case studies described in the previous section, the question of agonistic relationships is most clearly noted in relation to the See the Future project. In that project you have the city of Malmö as one important stakeholder, you have the design and innovation studio Unsworn Industries, you have us, but you also have the citizens of Malmö. If we initially take away the citizens from the discussion, it could be argued that the city of Malmö, Unsworn Industries and the research group involved have the same goal with the project: increasing the possibilities for Malmö citizens to have a say in how their city should be developed. But it is not that easy. Unsworn was paid by Malmö, and as always when you are a small company dealing with a big actor, discussions about payments are carried out in settings where one partner is much stronger than the other. There are also questions about rights to ideas and the physical products; if the project became successful, who would have the rights to export the idea to other cities?

However, the major question in this particular case concerns the citizens of Malmö. As previously described, the possibilities for citizens to take active part in the projects increased. But what happens after one has put great effort into participating? What is the end result? If you let a great number of citizens take part, you are bound to wind up with many different ideas and suggestions, and all cannot be used. It could also be the case that final decisions will go completely against some citizen suggestions.

This is a classic democracy problem, and there is no simple solution. But it has to do with accountability and with taking responsibility for one's actions. From the perspective of the researcher, this is a major ethical question. The way we deal with it initially is to think through very clearly which projects we really believe in, culturally, socially, and politically. Sometimes funding is available for projects that could be questionable in either of the above-mentioned aspects. Then we decline the invitation. But it is not enough that a project in itself looks promising. You have to look at the other stakeholders involved. And maybe most important of all, you have to take a long-term perspective. At some point in time, the funding for a project ends. But if you work in a concrete physical environment with many stakeholders, including the citizens belonging to the environment, the processes started will continue also after the end of the actual project. How do you deal with that? It is not a question you can deal with when you have reached that point. This is something you have to think about before getting involved.

The city of Malmö is relatively small, about 400,000 people. It used to be an industrial, working class city. But like many other Western cities, it has had to try to remake itself into a more knowledge-intensive city. It is also a very multi-cultural city, the most multi-cultural in Scandinavia, and it has neighborhoods with high levels of unemployment.

Many of our projects are based in areas that are often portrayed as problem areas. The citizens living there are for good reasons extremely tired of the portrayals of their neighborhoods in media, and they are often highly skeptical about new innovative projects trying to change conditions. They are also tired of researchers taking part in these projects, or coming to the neighborhoods in order to do interviews which may add to the negative portrayals.

Given this situation, we try to make sure that we do not start something that we cannot commit to working on for a long period of time. The researchers involved in the different neighborhood projects build trust by staying on. And we do not send students to work in areas where they are left on their own to do one-off projects. Instead we make sure their work becomes part of a long-term commitment.

Conclusion: The Role of the Media Researcher in a Changing Media Landscape

For some media researchers, the examples of applied media research that I have described in this chapter may not be considered as actual media research; with the exception of the Malmö City Symphony project, they do not deal with traditional media. You could also argue that in these cases, there is no interest in media as such; media are rather tools to achieve certain goals.

It is correct to state that the media are brought in here for specific reasons, and with specific purposes. But that does not mean that we are not interested in the qualities of each particular medium in itself. By using the media in these particular ways, we are able to explore the constraints and possibilities of each of them. In this sense, what we do is in line with the material turn that characterizes much media research today (Packer and Crofts Wiley 2011; Gillespie, Boczkowski, and Foot 2014). The work also adds new insights into what using the media could be about today, a use not restricted only to consumption. We thereby also move into contemporary discussions about prosumption (Ritzer 2014).

It is furthermore important to note that the notion of what a "medium" is today is not completely clear-cut (if it ever were). Taking the Internet of Things discourse seriously, most "things" may soon be regarded as media (Greengard 2015). This is not to say that all things should be regarded as media, but it is a discussion that media researchers should take a serious interest in.

That statement opens the question of what the role of a media researcher should be in this changing media landscape. I touched briefly on that question in the introduction to this article, stating that it is easier than ever to take both a collaborative and an interventionist stance. But it is also more important now than ever before.

Just a couple of weeks before writing this article, Oxford Dictionaries made "post-truth" into word of the year (2016). In such times, working actively with media, participating in actions, and starting processes becomes of utmost importance. And given the increasing complexity, not only of the media landscape but rather of the whole societal landscape, collaboration is necessary.

Coming to Malmö University from a traditional social science research environment, and getting the privilege of being able to work closely with design scholars, it has become clear how important for the media studies field theoretically informed, practice-based research can be. It leads to new knowledge on the properties and potentials of different kinds of media. It leads to knowledge on the uses of media. And it increases the possibilities for carrying out intervention-oriented work.

This does not mean that life as a researcher becomes easy—or easier. As discussed earlier, leaving the university buildings and taking part in processes in the wild means having to deal with accountability; playing an active part in an agonistic field of collaborative media means having to take responsibility for your standpoints and actions.

Working closely with people with other competencies and experiences also makes you aware of your own limitations. That could make you feel disheartened, at least initially. If you are used to working on your own project, with your own research questions, at your own desk, then working in a lab situation is different. But the point is not to become an expert on everything. You do what you are best at, the others do the same. But not in isolation! Working together, you learn from others, and others learn from you. Together you develop your project in ways difficult to foresee from the beginning. It is a challenge, but a rewarding one.

Notes

1 Vertov's movie is actually shot in a number of Russian cities, but is made to give the picture of just the one city.
2 Two PhD dissertations have focused on Medea's specific ways of conducting experimental media and design research, one with a focus on coproduction (Fischer 2015), and one on academic entrepreneurship, comparing Medea, MIT Senseable City Lab, and MetaLAB(at)Harvard (Simeone 2016).

References

Björgvinsson, Erling. (2007) *Socio-Material Mediations: Learning, Knowing and Self-Produced Media within Healthcare*, Karlskrona: Blekinge Institute of Technology. http://dspace.mah.se/handle/2043/5369

Björgvinsson, Erling, and Anders Høg Hansen. (2016) "City Symphony Malmö: The Spatial Politics of Non-Institutional Memory," *Journal of Media Practice* 17(2–3): 138–56.

Björgvinsson, Erling, Pelle Ehn, and Per-Anders Hillgren. (2012) "Agonistic Participatory Design: Working with Marginalised Social Movements," *CoDesign* 8(2–3): 127–44.

Dewey, John. (1916/1997) *Democracy and Education*, New York: The Free Press.

———. (1927/1991) *The Public and Its Problems*, Athens, OH: Swallow Press.

DiSalvo, Carl. (2012) *Adversarial Design*, Cambridge, MA: MIT Press.

Ehn, Pelle, Elisabet M. Nilsson, and Richard Topgaard, eds. (2014) *Making Futures: Marginal Notes on Innovation, Design, and Democracy*, Cambridge, MA: MIT Press.

Fischer, Josefine. (2015) *Knowledge Compromise(d)?: Ways and Values of Coproduction in Academia*, Lund: Lund University. https://lucris.lub.lu.se/ws/files/5491393/8084089.pdf

Gillespie, Tarleton, Pablo J. Boczkowski, and Kirsten A. Foot, eds. (2014) *Media Technologies. Essays on Communication, Materiality, and Society*, Cambridge, MA: MIT Press.

Greengard, Samuel. (2015) *The Internet of Things*, Cambridge, MA: MIT Press.

Hillgren, Per-Anders. (2006) *Ready-Made-Media-Actions: Lokal produktion och användning av audiovisuella medier inom hälso- och sjukvården*, Ronneby: Blekinge Institute of Technology. http://dspace.mah.se/handle/2043/3236

Huybrechts, Liesbeth, ed. (2014) *Participation is Risky: Approaches to Joint Creative Processes*, Amsterdam: Valiz.

Latour, Bruno. (2004) "Which Protocol for the New Collective Experiments?" www. bruno-latour.fr/sites/default/files/P-95-METHODS-EXPERIMENTS.pdf

Löwgren, Jonas, and Bo Reimer. (2013) *Collaborative Media: Production, Consumption, and Design Intervention*, Cambridge, MA: MIT Press.

Mouffe, Chantal. (2013) *Agonistics: Thinking the World Politically*, London: Verso.

———. (2000) *The Democratic Paradox*, London: Verso.

Oxford Dictionaries. (2016) "Word of the Year 2016 is Post-Truth." https://en. oxforddictionaries.com/word-of-the-year/word-of-the-year-2016

Packer, Jeremy, and Stephen B. Crofts Wiley, eds. (2011) *Communication Matters: Materialist Approaches to Media, Mobility and Networks*, London: Routledge.

Ritzer, George. (2014) "Prosumption: Evolution, Revolution, or Eternal Return of the Same?" *Journal of Consumer Culture* 14(1): 3–24.

Schön, Donald A. (1983) *The Reflective Practitioner: How Professionals Think in Action*, New York: Basic Books.

Simeone, Luca. (2016) "Design Moves: Translational Processes and Academic Entrepreneurship in Design Labs," Malmö: Malmö University. http://dspace.mah.se/handle/2043/21426

PART VI

Infrastructure

15

ARCHITECTURES OF SUSTAINABILITY

Kirsten Ostherr, Tara McPherson, Heidi Rae Cooley, Patrick Vonderau, Lisa Parks, Jason Farman, Elizabeth Losh, Eric Hoyt, Anne Balsamo, and Bo Reimer

Kirsten Ostherr

While doing applied media studies generally requires far fewer financial resources than doing applied science, issues of funding and sustainability nonetheless play a significant role in these projects. The multidisciplinary teams assembled to create applied media projects need space, technology, supplies, and human capital to succeed, and the pipelines for securing those resources are particularly limited in the humanities. This chapter asks contributors how they have managed to attain the needed resources for their projects, and what kinds of institutional homes they have found to house them. Further, to cultivate the necessary team members as participants come and go, we discuss what kind of background, training, and mentoring of undergraduate and graduate students, as well as faculty members, is needed to do this kind of work.

The analogy between applied media studies and applied science raises another interesting set of questions about funding, which has historically supported single-authored research in the humanities: What does it mean to be a humanities-based Principal Investigator? Successful competition for grants may look good on a curriculum vitae, but does involvement in applied media studies projects contribute positively, negatively, or neutrally to faculty advancement through promotion and tenure? What advice might the contributors offer to other scholars who are struggling to achieve recognition for the kinds of work they do beyond publishing in print-based media? What advice might contributors provide to university administrators on how to ease some of these challenges?

Contributors expressed a wide range of viewpoints on training, debating whether applied media practitioners need to learn how to code, with some questioning whether students need to be trained in applied media studies at all. At the same time, most respondents recognized that they had not been explicitly trained to perform the applied work they are doing now, nor to serve as "principal investigator," an uncommon structural role in the humanities. Participants recounted a similarly wide range of experiences with promotion, some noting that applied media studies engagements significantly slowed the pace of academic advancement, while others saw this work as an accelerant. All of the contributors emphasized the importance of funding, mostly to secure dedicated time for human participants, less to cover technology or physical space. Yet, the need for resources did not impede action; while recognizing the value of building sustainable infrastructure, the contributors also embraced a "boot-strapping" approach that fosters valuable kinds of creative, experimental practice.

How Have You Managed to Attain the Needed Resources for Your Projects, and What Kinds of Institutional Homes Have You Found to House Them?

Tara McPherson

As we initially launched the journal *Vectors*, we were fortunate to have the support of USC's Institute for Multimedia Literacy (IML). IML received a very large foundation grant that supported its programming for several years, and we were funded as one of their research projects in the later years of that grant. That luxury allowed us to assemble our team and to fund our early fellows. As we got established, we also began grant writing in earnest and haven't stopped since. We have been funded by the Mellon Foundation, the NEH, and the MacArthur Foundation. For many years, all of our funding was from soft money, and we continue to write grants, increasingly with other partners who are using Scalar for large projects.

We've also never had permanent space of our own. We shared the IML spaces as needed in the early years, especially for our summer workshops, but our team works in a highly distributed fashion on a day-to-day basis. We now share space with the Media Arts + Practice division in USC's cinema school and also have new relationships with USC's libraries. Our most important infrastructure is human—our core team and the various relationships we have built over time with libraries, archives, presses, and research centers—but this human network has to be tended to, both emotionally and financially!

Heidi Rae Cooley

The University of South Carolina is rich in resources that support our work, but we have no budget. So while:

> the histories that motivate the critical interactives we develop are local and the work to document them has been made available for mobilization;

we have ready access to relevant archival materials, as well as the commitment of the University Libraries' Moving Image Research Collections and Digital Collections, whose staff assist us with acquiring requisite materials;

  our colleagues generously offer their expertise and their time;

  the Critical Interactives course each year draws students whose energies produce impressive prototypes;

  we have a dedicated group of community members who are eager to share their stories for inclusion in the Ward One app;

  project development progresses intermittently, ceasing at the end of each spring term and only resuming at the start of the following spring.

We have received both internal and external grant funding (for *Ghosts of the Horseshoe*), with which we have been able to purchase equipment (laptops, iPads and iPods, and a 4-terabyte server, which is nearly end-of-life at five years old) and software licenses, as well as hire graduate and undergraduate students (fall 2012–spring 2014). Our projects "live" on our server housed in the College of Engineering and Computing.

Patrick Vonderau

All of my projects that I would consider "applied" have been financed through external funds. I managed to attain these resources by applying for them, often together with other colleagues. Although such funding always is institutionally administered, I wouldn't say that this work is possible only because of an institutional home. The opposite is true: in Sweden, applying for funds is a way to buy yourself out of your departmental duties, such as teaching or service. Ideally, of course, these projects would feed back into teaching and service. But teaching and service duties make for an enormous share of our workload, leaving little or no time for experimental or research projects. And the question of how to translate our individual work back into complex departmental cultures and curricula still needs to be answered. So far, I have designed one practice- and research-oriented course that engages student interns in a collective ethnography of media production. Establishing that course was rather difficult, not least because students need to commit extra time that isn't acknowledged in the credit points they are getting.

Lisa Parks

The funding resources supporting my collaborative research projects have ranged from European ministries of culture to the National Science Foundation to the U.S. State Department. Accessing this funding occurred by forming partnerships with other scholars and artists, developing project ideas together, and co-authoring grant proposals. Institutional homes for these projects have been in universities such as UC Santa Barbara's Department of Computer Science and the Center for Information Technology & Society, which I directed for three years. The center cultivates interdisciplinary research across the sciences, humanities, and social sciences. Some of my individual research has been supported by smaller campus grants at UCSB. After moving to MIT, I created the Global Media Technologies and Cultures lab as a site for collaborative research.

Elizabeth Losh

Right now I am in the midst of launching a new digital humanities (DH) initiative at my university, which is called "The Equality Lab." We are still struggling with things like logos and office furniture and catering for events, partly because I am still relatively new to the institution. I've done this in three ways: (1) partnering with social justice archives and inequality researchers thinking about launching DH projects who are looking at the university's painful legacies of slavery and segregation (the Lemon Project) and the exploitation of indigenous peoples (the Brafferton Project) and the region's regrettable histories of discrimination against LGBTIQ Virginians (the Mattachine Project); (2) partnering with campus initiatives for design, engineering, and innovation that are concerned about excluding women and people of color from participation; and (3) partnering with our office of eLearning and state OER initiatives, which have been thankfully not pursuing the misguided automation and depersonalization route popular elsewhere. I have seen the brochure for the capital campaign for the Equality Lab and hope that we'll eventually get the eight million dollars we need for permanent space.

Anne Balsamo

The AIDS Quilt Touch project has been supported by significant grants from the Office of the Digital Humanities at the National Endowment for the Humanities and from Microsoft Research. Several smaller institutional grants were given to participants from their respective institutions: University of Southern California, The New School, and the University of Iowa. The funding was always significant, but, in truth, never enough. The AQT project began several years before the NEH started the Digital Humanities Grants, which meant that the early work had been done without funding. The bulk of foundation funding ran out after three years. The project continues because it is now a *passion* project for members of the distributed research-design team. But their work will be uncompensated until the next grant is secured (which is not guaranteed ever).

Bo Reimer

The home for cross-disciplinary media projects has been the experimental lab Medea, set up in many ways in order to house such projects. Funding has come from Malmö University to set up the lab, but additional funding has had to be sought externally. Within Sweden we have acquired funding from The Swedish Research Council as well as from The Knowledge Foundation, which finances collaborations between universities and industry. We have also had a number of projects funded by the EU.

What Kind of Background, Training, or Mentoring Do You Think Is Needed to Help New Generations of Undergraduate and Graduate Students, As Well as Faculty Members, Do Applied Media Studies?

Tara McPherson

Graduate school officially trained me to work as a solo humanities scholar, producing single-authored books and articles. But, around the margins of graduate school, I developed another set of skills. As a union organizer and as a feminist activist, I learned a great deal about the rewards and challenges of collaboration and of crossing the town/gown divide. I think we are beginning to see the emergence of various programs that are trying to mainstream these skillsets within the academy. Some of this change is motivated by the changing nature of labor within the university, as workers increasingly have less job security. Incorporating team-based work with museums, non-profits, archives, and other partners helps graduate programs prepare students for a greater variety of careers. But the change is also motivated by the recognition that many pressing social problems do not neatly fit the contours of a single discipline. Students needs opportunities to work as a member of a team, to do research across disciplinary boundaries, and to produce outputs that are not just text-based articles and monographs.

At my own university, I've been involved in a few efforts to foster such opportunities. We regularly involve students in the work of our lab, and the relationship is mutually beneficial. We learn as much from the students as they learn from us. That sounds like a cliché, and it is. But it's also true. Students often approach software design and scholarship from different angles than faculty do. The insights they bring are very helpful in our development process.

We are also building programs and institutes to support these different modes of learning and working across the university. I direct the Sidney Harman Academy for Polymathic Studies, a space for students to come together to collaborate and think together outside of the boundaries of a single discipline.

The Academy is largely an opt-in, non-credit space, and we find that this struc-ture attracts USC's most interesting and intellectually curious students. Many of the students, and especially the undergrads, express frustration at the bound-aries created by the structure of majors and minors on campus. The Academy allows them to engage in lively dialogue with students from other fields, to plan conferences and events together, and to undertake mentored research projects. At the graduate level, I am lucky to be affiliated with MAP, the practice-based Ph.D. program in the cinema school. This program is very unique within the U.S., as it allows humanities, social science, and arts students to create hybrid dissertations that value practice equally with theory. These students often work in faculty research labs (or in their own labs!) while also learning humanistic modes of inquiry. Europe had such hybrid programs before the U.S., but, as I travel around the country, I see more and more interest in creating similar structures that exceed traditional disciplinary boundaries.

Heidi Rae Cooley

For the record, I am less an advocate of encouraging humanities students to learn Python or Unity, for example, than I am of getting them to speak thoughtfully and effectively with those who do know programming languages. As such, I focus on cultivating productive communication across diverse modes of thinking. Of course, I do not discourage students from dabbling in the more technical side of project development. In fact, I think that productive cross-disciplinary conversation often results in really pointed forays into tech-nical experimentation.

Fostering a new generation of applied media studies scholar-practitioners begins with modeling good collaborative practices and opening up the collab-orative process to newcomers, inviting them to participate in the process in meaningful (not simply functional or peripheral) ways. This requires establish-ing and sustaining an intellectual community wherein students—undergraduate and graduate—and interested others feel themselves to be members contributing concretely to a larger whole. Investigation and experimentation are cooperative and conversational endeavors. Faculty and students alike need to engage actively and according to their disciplinary strengths in discussions about goals/outcomes of a project so that the vision is shared and can develop over time as the compo-sition of the group changes, funding surfaces/recedes, and priorities shift.

I think it productive to frame "applied" work as belonging to the collective. No single individual (e.g., researcher) "owns" the project, even as one person may be designated the lead or project investigator. All members of the team are stakeholders. In this regard, I think it particularly important that students who are funded to work on a particular project understand that they work for the project, not for a particular faculty member. By "for," I mean "in the interest of," that is, they are responsible to the project and its vision as conceived by the whole team. It really should be about evolving an ethos, that is, a way of thinking and practicing intellectual labor, not as personal, isolated endeavor, but as a communal process of exchange and growth, wherein individual scholarly strengths come together to "solve" a shared "problem." This takes time and effort—time reading together across a variety of fields of expertise; time in conversation, learning to speak across traditionally-held and/or -imposed disciplinary divides; time brainstorming, concept-mapping, wire-framing (whatever affords a shared sense of the "big picture"); time dedicated to the iterative process of building and testing and revising; time cultivating trust.

Patrick Vonderau

There are two issues here about which I disagree, given my own situation and experience working in Sweden and Germany. The first is the idea that doing applied media work also would necessitate fostering a generation of scholars doing that very same work. This presupposes that "applied media studies" is indeed a discipline or a field where these students really can have a career. But there is very little indication that this is the case, at least when considering university employments in Europe. The same even holds true for the digital humanities. Although there are a few DH labs here and there, and although young scholars working with digital tools or other inventive methods may sometimes have a career advantage, this kind of work hardly leads to more or better career opportunities, at least for now.

The second is the idea that we would need to translate our own experimental and often informal, ad hoc "applied" research practice into some sort of formal training. Why? This seems a bit premature to me. It is still too early to say if our so-called digital tools or methods will stand the test of time. Do all humanities scholars now have to learn programming languages such as Python? Or, to paraphrase Clifford Geertz, studying cannibals, do we have to become cannibals? I don't think so. Claims like these reiterate older ones of all film students absolutely having to learn handling a camera or editing table. For some, these turn out to be life-changing experiences, but many others may prefer a technical division of labor and the lively exchange with, say, professional coders or film-makers. Future generations of students will come equipped with programming and video production knowledge anyway. So we rather might think of how to teach humanities' way of knowledge production to those trained in the standards and practices of other fields.

Lisa Parks

This is a complex question as it depends on how one defines applied media studies. In general, I think it requires openness to other disciplines, an interest in engagement beyond the university setting, and creativity with regard to research questions and methodologies and the presentation of research findings. It is helpful for participants in applied media studies to know how to communicate across different contexts and fields, and to be able to produce media, whether a short video or social media stream, an infographic, or a blog. Some might say it helps to have training in areas as diverse as ethnography, data visualization, and coding. With regard to mentoring, it is helpful to involve undergraduate and graduate students in collaborative research projects and to reflect with them on the process of collaboration so that they can acquire a sense of what is possible and what works, and what may not be possible or not work. This enables students (and faculty) to establish realistic expectations regarding project outcomes. A lot can happen through regular meetings attended by a group of people committed to working together. Projects get going when institutions create and normalize spaces for interactions and gatherings across disciplines.

Jason Farman

The only requirement to do this kind of work is that a person have curiosity. I have been self-trained with all of the technologies that I use for my hands-on work. I've never taken a formal class in design or coding. One of the most important lessons I pass along to my students is that they can teach themselves how to use these tools. If my students have the confidence to pursue their curiosity and teach themselves how to use these technologies for applied media studies, they have overcome a massive hurdle for this field. So, from my perspective, the important lessons that a student must learn are: (1) you can teach yourself the tools needed for the kind of work you hope to do and, (2) you will continue to teach yourself, because the notion of mastery is illusory and we must instead embrace lifelong learning. There are so many rich resources for learning how to design, code, and engage in other modes of applied media studies that students can confidently experiment with these tools and teach themselves how to use them. As my students do this, I encourage them to contribute to the incredible body of work, sharing the knowledge they gain with others. Alongside becoming familiar with new tools (and getting comfortable with new interfaces), it's important to teach students how to critically approach them. Most of the tutorials that teach someone how to use one of these tools will not teach them how to employ the tool in a critical way or to answer complex questions that our field wrestles with. So, it's important that the student learn the basics of employing a tool, while simultaneously learning to ask the right questions about that tool (and the right questions *with* that tool).

Elizabeth Losh

I know this is a cliché but I do actually think I learn more from graduate students and pre-tenure faculty about how to connect a rich life lived on social media with new ways of thinking about scholarship that occurs in places other than the single authored monograph from an academic press. In the humanities I think having interdisciplinary research cohorts can be a good way to find mentors in applied media studies, and senior scholars in FemTechNet like Lisa Nakamura, Anne Balsamo, and Alexandra Juhasz have been incredibly generous to those who are in precarious positions in the academy or those who are being coerced to replicate the closed modes of scholarship traditionally associated with the academy.

Eric Hoyt

So many people doing applied media studies work would benefit from greater computational literacy—from a better understanding of how algorithms, databases, and the things we call "digital" actually work. It's much easier to get a project started if you can just roll up your sleeves and start coding rather than writing a grant to hire someone next year to do the coding for you. And, in collaborating with programmers and engineers, computational literacy is essential for being able to communicate and work together toward a common goal. There is a steep learning curve when you first try your hand at coding. So it would be great if we could support graduate students and professionals in our field with time and training to complete this learning curve.

To be clear, the end goal isn't to transform all media studies scholars into expert programmers and engineers, ready to step into cubicles in Silicon Valley. The goal of computational literacy is to empower scholars to get projects off the ground, effectively collaborate in teams, and develop a greater appreciation of the thought and labor that goes into digital projects.

Anne Balsamo

I have argued that people working on applied media studies projects need a foundation in theories of technology, especially to understand the reproductive logics of technology development. They need to understand culture is both a resource and an outcome of the development of any media project. They also need training in design skills and methods, especially to understand approaches to human-centered design. Mentoring guidance should focus on how to assemble cross-disciplinary teams, and then how to cultivate collaboration across differences (that arise from different disciplinary backgrounds).

Bo Reimer

It is crucial for a new generation of applied media studies scholars to both acquire specialist knowledge within a particular part of the media studies field, and to acquire the skill of being able to contextualize that knowledge, both within the media studies field and more broadly. He or she has to be able to be an expert on something as well as being able to understand how the things he or she is an expert in functions in relation to other things. For this to happen, it is important that the university offers both undergrad and master's programs that take this dual view on media studies. Preferably such programs would collaborate with programs in, for instance, interaction design, media technology, computer science, or the arts. Mentoring is important, and the best way of mentoring would be within the framework of projects carried out in the wild.

The Analogy with Applied Science Raises Another Interesting Set of Questions About Funding, Which Has Historically Supported Single-Authored Research for Humanists: What Does It Mean to be a Humanities-Based Principal Investigator?

Tara McPherson

Being a PI in the humanities means learning new skill sets (such as writing and managing grants or working in a diverse team). But, perhaps more importantly, it also means finding ways to explain what you are doing to colleagues who do more typical modes of humanities scholarship. At a practical level, promotion committees and department chairs rarely know how to evaluate or make sense of collaborative, team-based research. Scholars who enter into this work should know that it comes with risks. Your career might not be quite legible to many of your peers and may be met with some skepticism. Of course, these risks are counterbalanced by the rewards of collaboration across and outside of the university.

Heidi Rae Cooley

Being a Principal Investigator means being an effective project manager. Being an effective project manager means essentially three things. One, an effective project manager attends to bureaucratic tasks with efficiency and care. She manages whatever budget exists, makes sure deadlines are met, oversees the productivity of any hired labor, and schedules and plans meetings of the project team. She does not do all of this work by herself, but she does agree to ensure that the business of the project is conducted responsibly and in a timely fashion. Two, an effective project manager accepts responsibility for guiding the process of project development. Beyond scheduling and planning meetings, an effective project manager functions as a facilitator, cultivating an intellectual community and facilitating the kind of exchanges that ensure that project development progresses. She challenges team members to take risks, to stretch themselves; at the same time, she provides inspiration and vision. Three, an effective project manager is responsible to and respectful of the team of collaborators. She strives to listen actively and responsively to individual and shared ideas, questions, and concerns, at the same time she works to ensure that all voices can be and are heard. In this regard, she is watchful for (un)intentionally imposed silences and acts promptly to intervene in counterproductive dynamics caused by miscommunication, feelings of exclusion, moments of aggressiveness, and so on.

Patrick Vonderau

Here, the situation may be different in the U.S. In Sweden or Germany, it is not uncommon to apply for collaborative research projects even in the humanities.

Lisa Parks

A humanities-based PI might be a single person working on a project by herself, an equal research partner in a larger interdisciplinary collaboration, or a key bridge builder and liaison between the university and a public entity or commercial organization. It is important that humanities agendas continue to thrive and proliferate in and beyond the university setting. Silo-ization has not served the humanities well. It is incumbent on humanities scholars—PIs— to get out there, figure out how to communicate what they do to those in other parts of campus and beyond, and imagine ways of working together without feeling as if they are "selling out." The university is a dynamic setting and humanities scholars can continue rich legacies in their disciplines while expanding and reinventing these disciplines as well.

Elizabeth Losh

The work of Lisa Nakamura, Sharon Irish, T.L. Cowan, and many others about the importance of community managers rather than principle investigators is important in highlighting the value of labor that is often affective, feminized, and invisible. In the field of digital humanities, Sharon Leon has been writing some very insightful work about why the principle investigator model excludes librarians, alt-ac people, and many important constituencies. As someone who works a lot with collectives that are intra-academic but not part of formal professional organizations (although we do have meet-ups at conferences, often with the assistance of groups like HASTAC), I know that getting funding for community management is a constant headache.

Anne Balsamo

As a Principal Investigator, I am responsible for assembling the team to make sure that the range of skills and expertise is available for the development of the project. But it is also my task to guide participants, theoretical understanding of what it means to "make a project"—as I've written elsewhere—the project does not precede the interactions and improvisation of those who make it manifest. Being a PI requires the skills to translate among discourses, to guide negotiations among participants, and to cultivate the "stakes" for each. Cross disciplinary team management—the responsibility of the PI—requires understanding of what is on offer for each person in terms of disciplinary insight, research, or scholarship.

Bo Reimer

In our projects, the notion of a Principal Investigator has not been especially important.

Has Involvement in "Applied" Media Studies Projects Contributed Positively, Negatively, or Neutrally to Your Advancement As a Faculty Member through Promotion and Tenure?

Tara McPherson

The work I have done in applied media has negatively impacted my timeline for promotions. It has also negatively impacted the careers of scholars I work alongside. If I had chosen to pursue a 'normal' path as a humanities scholar, I would have moved rank from Associate to Full Professor more quickly, because I would not have needed to learn new skills nor would my colleagues have been perplexed by my work. While my dean was generally supportive, my division colleagues were not really able to evaluate my work in meaningful ways. Nonetheless, I don't regret the path that I took. My work as a media scholar has been deeply enriched by being closely involved with media production.

I have also reviewed many tenure files for junior scholars who work in various types of applied media. They often produce 150–200% of the work of their peers, creating digital or other media projects while also authoring monographs of their own. This practice is deeply unfair. I feel it is slowly beginning to change but not quickly enough.

Heidi Rae Cooley

The "applied" component of my research has been fundamental to my advancement to the level of Associate Professor. My first book, *Finding Augusta: Habits of Mobility and Governance in the Digital Era*, with its digital supplement, *Augusta App*, was the cornerstone of my tenure and promotion case. And the majority of my other publications cite my "applied" work, using them as examples for thinking about the kind of habit-change I argue is possible—and productive—in the mobile, connected present. Moreover, several of my more recent publications have been co-authored with my collaborator, Duncan Buell; I see them as extensions of the "applied" work we do together.

Patrick Vonderau

Having received funding for projects and participating in such projects was definitely key for my advancement as a faculty member, especially when advancing from associate to full professor.

Lisa Parks

It depends on where one is situated. The UC system is quite open to research that is experimental, work that is field expanding, and work that engages with publics. The Film and Media Studies department at UCSB, where I worked for eighteen years, was very open to the idea that research findings can be published or presented in different ways, whether in books published by university presses or op-ed pieces, peer-reviewed articles or art exhibitions. At the end of the day, though, there is definitely a higher value placed on traditional modes of publishing in the humanities in the upper levels of administration and review. In research institutions, the monograph reigns supreme in the humanities, but baby steps are being made and new book-length projects with experimental formats are emerging and qualifying people for tenure and other promotions. It is important for those in applied media studies to solicit external reviewers who "get it," so to speak, as these reviewers can explicate the significance and impact of applied media research for internal review committees.

Jason Farman

During my process of going up for tenure, there were places where this kind of work was discussed in my own statement, on my CV, and in the committee's report on my work. The applied work that I did—whether that be app development, performance art, or graphic and web design—never tipped the scales in any direction during the process. While the university created structures that seemed to recognize this work, as noted in the categories on the official university CV template and in my department's evaluation of my work, the real value was placed upon traditional scholarly publications such as books and peer-reviewed journal articles.

In my own work I advocate for pathways that allow the work done in applied media studies to translate across multiple venues, genres, and modes of scholarly output. For me, designing a location-based story app was an incredible process that brought me together with my students and members of our community to attach narratives to locations that highlighted the often untold stories about those locations. I then translated this work into peer-reviewed journal articles and an edited book called *The Mobile Story: Narrative Practices with Locative Technologies.* Those articles and my edited volume were weighted differently than the hands-on work of creating a location-based story app. Regardless, I don't believe I could have produced that traditional scholarly work without having engaged in the creation of this kind of mobile story app. Though the hands-on work was not recognized in a way that matched the amount of effort and innovation involved in the story app that we created, this work unlocked ways of knowing that allowed me to produce compelling scholarly articles and books. Until the scholarly review process catches up with thinking of applied media studies as a rigorous mode of scholarly output, I encourage people to find ways to articulate their findings across multiple media and multiple modes of scholarly engagement, some of which will be recognized during the promotion and tenure process, and some of which will not.

Elizabeth Losh

I think these activities often look strange on a CV, particularly when there is also external press coverage involved that might distort what a faculty member might be trying to do in thinking about how communities use (and are perhaps used by) particular media technologies. For example, people affiliated with the Selfie Course have often had their words taken out of context by the media, which might create some distortions in the online results when review committees Google people (as they inevitably do) to place a CV in its larger rhetorical context. In my own case I think these activities have been beneficial because my alliances with collectives in applied media studies have encouraged me to explore research questions that I might not otherwise have considered. I was also able to translate a lot of the intellectual work that I have done as a blogger, both on my Virtualpolitik blog and as a guest blogger at DML Central and the Institute for Money, Technology, and Financial Inclusion, into writing for peer-reviewed publications. This translation means that I am already thinking about how to incorporate interviews, field work, archival work, and visual documentation into the framework of scholarly criticism.

Anne Balsamo

There is no way to know for sure if my ability to secure grant funding for my projects has actually had an impact on my professional advancement. I suspect that to outside evaluators this is considered an important part of my portfolio of skills. I know that when I look at a candidate's CV (for a faculty position), I make note of the grant history. Even small internal grants make an impact because it suggests the ability to conceptualize a project and bring it forward into some degree of realization.

Bo Reimer

For me, involvement in applied media studies projects has not been regarded as something negative. It is, however, crucial to make sure that these projects lead to publications or other forms of meaningful results.

What Advice Might You Offer to Scholars Who Are Struggling to Achieve Recognition for the Kinds of Work They Do beyond Publishing in Print-Based Media? What Advice Might You Provide to University Administrators on How to Ease Some of These Challenges?

Tara McPherson

Seek out the mentorship of like-minded senior colleagues outside of your department and university. These scholars can help you find ways to explain your research and applied activities to your colleagues. Likewise, build a support network among your peers to share resources, ideas, horror stories, and successes. Become familiar with the guidelines for digital and public scholarship offered by various societies like the MLA, CAA, and the AHA. Find out if your university has such guidelines and look at the guidelines at other universities. (For instance, USC has recently added guidelines for both digital and collaborative scholarship to its faculty handbook.) Share these guidelines with your chair and administrators.

If you are offered an academic job, initiate discussions about expectations for tenure or other promotions before you sign a contract. Will a book be required for advancement or will public or applied work be accepted? What standards of evaluation will be used? Who will be available to mentor you? I worry that we are producing a wonderful cohort of hybrid scholars at USC who will be hired into departments that don't know what to do with them. Senior scholars engaged in applied research have a deep ethical responsibility to help shift expectations around what counts as research today.

University administrators have a responsibility to make sure emerging forms of research are fairly and thoroughly evaluated. They should be familiar with the guidelines that learned societies have developed and that their peer institutions are implementing. They should articulate their expectations clearly to new hires and commit to these expectations in writing. They should ensure that new faculty and staff have mentors who can support their work. Finally, they should address issues of labor that extend beyond tenure-stream faculty, making sure that staff who work on hybrid or applied projects have good working conditions and receive proper credit for the work that they do.

Heidi Rae Cooley

To achieve recognition, one must have visibility. One way to accomplish this is to publish about the "applied" work in print-based media. I think the two modes of scholarly practice are mutually informing and can be—perhaps ought to be—mutually constitutive.

Upper-level university administrators should want to support collaborative, "applied" media studies projects. And they are, judging by the increasing number of centers, institutes, initiatives, etc., dedicated to this kind of scholarship. However, I think it paramount that the establishment of such entities be accompanied by clearly stated policies and procedures that result in reasonable and equitable structures for: (1) defining the unit's vision, mission, objectives, and method of self-assessment, (2) appointing and evaluating persons in leadership positions, (3) defining the role of any group serving in an advisory capacity, (4) ensuring the unit's openness to several varieties of scholar, and (5) promoting productive, respectful collaboration among faculty and students (and community members).

Directors of programs and chairs of departments should support faculty who have succeeded in establishing collaborative relationships with colleagues in other disciplines and are eager to team-teach courses that foster critical awareness of historical/social/political importance through "applied" studies. This may result in lower enrollments for each instructor; it has for my collaborator Duncan Buell and myself. But, we've found ways to offset the "imbalance." First, the success of our and our students' efforts, as demonstrated at end of term public presentations-demonstrations (five as of spring 2016), has translated into greater visibility—institutional, as well as within the broader community—for our respective disciplinary units. Second, we each regularly teach larger lecture classes for our respective departments/programs.

Patrick Vonderau

Again, the situation might be different for scholars in Europe. I have never heard of colleagues who successfully engaged in forms of applied media work and really struggled to achieve recognition for that work. To the contrary, such scholars very often are seen as representing a new class of knowledge workers, distinguished from what consequently appears as the grey, undifferentiated mass of regular humanists. Being skilled to use digital tools of whatever kind, or to engage in socially or politically relevant work is definitely acknowledged. Having said this, I do not think that applied media scholars need extra compensation (apart from the recognition, grants, or awards they already get), or that traditionally working humanists should be regarded as outmoded. We have to be careful not to think in terms of New Public Management or economic administrators now in charge of monitoring our resources and skills.

I am not sure university administrators are particularly open for advice, given their own tasks and dependencies. Here is one, however: Quantification doesn't help. But since even humanists now tend to think that number counting is a game-changer, that hardly would come off as a convincing argument from a humanist.

Lisa Parks

In order to maintain rigorous standards, university administrators need to continuously educate themselves about the changing research in humanities fields. Researchers seem to be way ahead in terms of their experimentation, and administrators are often holding on to old standards that were in place when they first entered higher education. If I were an administrator I would schedule meetings with groups of newly hired faculty in the humanities and ask them about the current conditions in their fields, explore whether the institution meets their ambitions, and figure out ways to best support their research. I would also meet with chairs of humanities departments and emphasize the need to recognize and reward emergent and experimental forms of scholarship that demonstrate rigor, creativity, interdisciplinarity, field expansion, and public engagement.

Jason Farman

Advice that I gave to others even before I received tenure is the same advice I give post-tenure: don't let what is or is not recognized in the tenure process dictate the kinds of work you do. Whether something is or isn't recognized in the tenure process should not be a reason we pursue something or decline our involvement with it. Sometimes tenure makes us value the wrong kinds of things because a very narrow set of practices are recognized as leading to successful tenure reviews. On one hand, there is a form of recognition that is provided in the tenure and promotion process; on the other hand, there is the recognition that we get from our students, our colleagues, and the broader public who might be impacted by the work that we do. While the tenure and promotion process may be slow to adequately recognize work that scholars do in applied media studies, there are many avenues for recognition outside of the sphere of tenure and promotion. It's important that we as individuals, scholars, and activists allow ourselves to appropriately value those other modes of recognition as well.

University administrators need to have close contact with researchers in their divisions not only about the future of this kind of scholarly work, they should also be involved *during* the process of creating this work. Administrators should understand how this work gets created, not simply review the finished product as a step in the promotion and tenure process. Much of the work that we do is not simply about a final product; it is often just as important to understand how the work was made. If making is a mode of thinking, then *the process is an essential part of how we produce new knowledge*. Finding ways for universities not only to recognize scholarly output but also to recognize *scholarly process* will be a major challenge in the years ahead.

Elizabeth Losh

This is going to sound jaded, but I think at this point getting recognition for non-print based work is an uphill battle. To get their ideas into print I encourage people to explore collaborative authorship, provide more curatorial text to digital humanities projects, beef up artist statements into articles, and network with editors who want to launch peer-reviewed publications that might be more adventurous. I earned tenure on my books, which I think will continue to be the case in the humanities for a long time (as Diane Harley's research confirms).

It's easier if these initiatives are considered foundational to an institution and if there is historical memory tied to origin stories about faculty buy-in. When I directed the Culture, Art, and Technology Program at UC San Diego, I was able to take advantage of the fact that the program was part of the founding of a residential college devoted to twenty-first-century initiatives (science and technology studies, multimodal publishing, practicum experiences, coding literacies, public rhetoric, etc.) and its ambitious (and often underfunded) academic plan had been voted on by faculty and sustained with a continuing contingent of about 200 faculty affiliated with the college.

Eric Hoyt

My advice: don't give up, but do calibrate your expectations. You will probably need to keep publishing traditional peer reviewed books and articles. This does not need to be viewed as a burden. Writing and publishing remain great ways to communicate knowledge and express complex arguments.

Applied and digital projects will probably give you a less prestigious credential than a university press book. But a digital project can give you things that are equally valuable and even more satisfying—visibility and goodwill among other scholars in your field, for example, and invitations to speak at events and participate in projects all around the world. This has been my experience in working on the MHDL, Lantern, and Arclight. I wouldn't trade the time I've put into them for anything.

And, remember, it's not all about you and your career. Applied media studies, by its nature, encourages you to develop projects that add value to the lives of other people. If your applied work is making a difference and improving people's lives in some way, then that's something to celebrate and keep doing more of.

I'm grateful to work at an institution, the University of Wisconsin-Madison, that explicitly values outreach and the production of work that positively impacts lives beyond the ivory tower. We call this "the Wisconsin Idea," but no state or university holds a monopoly on public service. It would be great to see more universities implement promotion criteria and funding streams that reward faculty, staff, and grad students for pursuing public-facing, applied humanities work.

Anne Balsamo

I am struck by the realization that everyone is writing (digital work), but no one is reading (it). There is so much produced now, in the form of historical place-specific apps, multimodal essays, online publications, etc. The creation of digital works of scholarship is not as challenging now as it was in the early days of computational humanities. The real challenge, I believe, is finding a way to get people to actually read or use your application or digital work. I would highly advise a younger scholar to pay attention to scaffolding the use of their project, to suggest and promote ways for it to be integrated into syllabi for example. Just building it in no way guarantees an impact or an audience.

Now that I'm in the position of a senior administrator who has responsibility for nurturing the careers of junior faculty, I am committed to providing some level of resources for project prototyping. The first level of prototyping of an applied media studies project can be with paper and printout; detailed story boards, speculative design fictions, improvisational enactments. The idea of the project has to be materially prototyped; this need not require much funding, but it does require space and time (and maybe also student project assistants). My advice: small pots of money, project development space, course release for prototyping and grant preparation, and funds for the employment of student project assistants.

Bo Reimer

It is still very difficult to achieve recognition as an academic without publishing in traditional fora. But if you combine such publishing with other kinds of outputs then things will be fine. In Sweden right now, there is a move towards finding relevant indicators on how academic work impacts on the surrounding society. Such a move could hopefully be of value for applied media approaches.

It would be good if university administrators could assist in constructing valid indicators on the value of applied media approaches.

16

BUILDING A LANTERN AND KEEPING IT BURNING

Eric Hoyt

It's 3:27 p.m. on what I had been telling myself would be a highly productive writing day. But instead of typing out paragraphs of prose, I have spent the day responding to testy e-mails, Googling error messages, and working with Pete to get Lantern back online.

The website crashed. Again.

The crash happened sometime early in the morning. Our usual solution—a reboot of the index's virtual server—is not working. And so Pete and I are spending a day that we should be using to make progress on our new projects just trying to get an old project to work. Even as we succeed in restoring service (changing the virtual server's RAM allocations did the trick), I am aware of the fact that we will be fixing Lantern again soon, and that I remain behind on all of that writing I was going to do.

I open my essay with this example because it exemplifies something I've learned about working on digital projects: you are never really finished with them. Software needs to be upgraded, new information needs to be added, and servers go offline. The work of sustaining a digital project can take up more time than creating the project. And it demands your attention and time in a way that is less predictable and controllable. The results can be stressful, but these glitchy moments can also be valuable reminders that there are people out there, right this minute, who are using and depending upon your work. Sometimes, it takes something going wrong to hear from them and remember why you put in all the work in the first place.

This essay is a personal account of Lantern, spanning the process of development, the public launch, and the past three years of maintenance and growth. The personal aspect of this story is important to stress: I have tried to capture the emotional highs and lows of this work, and the occasional haphazardness, rather than presenting everything in the clinical prose of a funding agency report. Since this is a personal essay, we should also indemnify up front my many collaborators on Lantern and its larger project, Media History Digital Library (MHDL). I feel like I should invoke here the language that appears at the beginning of DVD commentaries: the views expressed here may not represent the views of the larger organization. If anything, I suspect readers will feel sympathy for the people who have to put up with my idiosyncrasies and baseball metaphors on a daily basis.

I have structured this essay across four sections that span the life of this particular project. My hope is that these phases are familiar and applicable to other developers of applied media studies or digital humanities projects as well:

1. Addressing a Need
2. Starting Up on a Shoestring
3. Completing and Launching the Software
4. Sustaining and Improving the Project

If there's a theme to this essay, it's the importance of a special kind of team dynamic that building and sustaining Lantern required. Lantern would never have advanced from one stage to the next without the input and collaboration of tremendously talented people. But it also would never have advanced without me playing a very direct and active role—rolling up my sleeves, and, in some cases quite literally getting my hands dirty.

Addressing a Need

Lantern (http://lantern.mediahist.org) is first and foremost a search platform. So before discussing how the project was developed, it is important to understand the collection that it searches, the MHDL (http://mediahistoryproject.org). All applied media studies projects should be able to articulate the need and audience they are serving. Otherwise, you risk investing a tremendous amount of time into a PFN (project for no one). In this particular case study, understanding the need for Lantern requires understanding the need for a search tool to access MHDL.

The MHDL emerged to serve a specific need and base of users. For decades, researchers relied on old trade papers and fan magazines to write the histories of film, media, and broadcasting. Industry trade papers, such as *Variety* and *Film Daily*, offer extensive documentation about the development of the media industries—chronicling the paths taken, the promises never fulfilled. For historians of silent cinema, reading through several years' worth of *Moving Picture World* on microfilm became a rite of passage. You put in an inter-library loan request for several microfilm reels. Once they arrive, you reserve a machine, carefully load a reel, and slowly crank, looking page-by-page through this important industry publication in search of articles and advertisements pertinent to your research topic.

We all knew there were ways of using digital technology to improve this research process, but film historian David Pierce was one of the first to act on it. He knew the breadth of publications that existed, and he knew the institutions and private collectors who possessed original copies (not simply the microfilm). He understood that books and magazines could be digitized for a reasonable price at scanning centers managed by the Internet Archive. David also possessed twenty-five years of experience investigating the copyright status of books and films. Whereas most Hollywood feature films from the 1920s through 1950s were still under copyright protection, most of the trade papers and fan magazines of this period are in the public domain. As the adage goes, "yesterday's news wraps today's fish." The Hollywood studios applied for renewals for their copyrighted content, but the news publishers who covered the studios did not.

In 2009, David founded the MHDL and coordinated the digitization of ten volumes of the fan magazine *Photoplay* from the collection of the Pacific Film Archive. Setting the precedent for everything subsequently digitized by the MHDL, the scanned magazine issues became freely available online for broad public access (see Figure 16.1). In 2011, the MHDL dramatically scaled up its scanning activities, digitizing over 200,000 pages of magazines and books from several private collections. I was living in Los Angeles at the time, and I had gotten to know David through my dissertation research on the history of film libraries. I was eager to contribute to the mission of the MHDL, and I was able to get involved as the project's digitization coordinator—driving to the homes of LA collectors, packing up magazines from their collections, delivering them to the San Francisco Internet Archive scanning center (stopping by In-N-Out for a burger along the way), and ultimately returning the magazines back to the spaces waiting for them on their owners' shelves.

The MHDL's scanned books and magazines appeared on the Internet Archive's website, but they were difficult to find among the Internet Archive's millions of other digitized items. We needed a website that represented the MHDL's identity, allowed for easy browse-ability of the collections, and enabled users to quickly find what they were looking for. I collaborated on building the website with Wendy Hagenmaier, my brilliant sister-in-law who was completing her master's degree at the University of Texas at Austin's School of Information.

In designing the MHDL website (built in WordPress), Wendy and I tried to think carefully about how our user-base of scholars, fans, and students would explore the site. The resulting website—which contained sub-collections devoted to the Hollywood Studio System, Fan Magazines, and Broadcasting History, as well as collapsible chronological lists of magazines—was most intuitive for those scholars and fans who were already familiar with the publications featured on the site. These knowledgeable historians and fans already knew the difference between *Film Daily* (a trade paper) and *Photoplay* (a fan magazine).

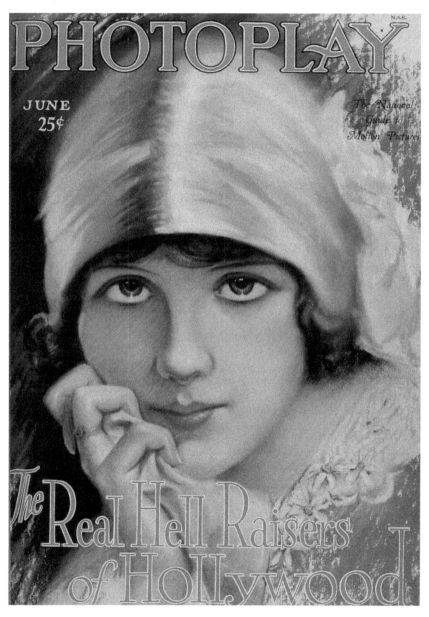

FIGURE 16.1 This June 1927 issue of *Photoplay* was one of the first magazines digitized by the Media History Digital Library. It was scanned from the collection of the Pacific Film Archive, funded by an anonymous donation in memory of Carolyn Hauer. http://lantern.mediahist.org/catalog/photoplay3133movi_0741

They also had the years of film releases burned into their brains; they knew to look in the 1939 *Film Daily* issues if they were looking for information about the premiere of *Gone with the Wind* or *Mr. Smith Goes to Washington.*

However, we had another, larger potential group of users who had no experience researching historic industry periodicals. For this group of users—which included most undergraduate film students and the fan base of cable channels like Turner Classic Movies—the initial MHDL site was challenging to navigate. You knew you wanted to see historic news coverage about Cary Grant, but you didn't know where to start looking. Should you start browsing in the Hollywood Studio System Collection or the Fan Magazine Collection? And what years would be the best to look at? Why could you perform keyword searches within an individual volume, but not across multiple volumes at once?

To better serve expert and non-expert users alike, we began developing Lantern, a full-text search platform for the MHDL's collections. Why call the platform Lantern and not simply "search"? The latter option probably would have avoided some brand confusion; we occasionally hear from users who mistakenly think that the MHDL and Lantern are unrelated. But I was getting ready to apply for academic jobs, and I thought it would sound more impressive to say that I was building a platform called Lantern (which would undoubtedly take a couple of months to complete), rather than simply calling it a search box. Toward that end, "Lantern" was a good name to choose. My miscalculation was in assuming that the project's development would be measured in months and not years.

Starting Up on a Shoestring

We had a goal, and we had a name. Beyond those things, Lantern had the following assets and liabilities at its inception. The assets were the MHDL's growing collection of magazines, hosted by the Internet Archive; a fast, open source search engine (Apache Solr) that we could use to create a search index; some personal knowledge of web design and computer programming, plus the curiosity and drive to learn more; close proximity to an excellent computer programmer (i.e., my father-in-law, Carl Hagenmaier). The liabilities were no project funding; just enough design and programming knowledge to bite off more than I could chew and get myself into trouble; and limited time: in September 2011, I had six-month old twins and a dissertation to finish, plus I was applying for academic jobs.

Fortunately, as it turned out, the asset of Carl Hagenmaier more than offset the liability of my ignorance. I would spend six hours getting myself into a jam, then Carl would help me solve the problem in a matter of minutes. I would probably still be lost in some Unix command line terminal without him. Wendy also made major contributions at this stage—helping to structure the metadata in a consistent and logical manner that allowed it to be indexed in the search engine. How did I get so lucky as to marry into this family?

Carl, Wendy, and I were able to set up a search index of the MHDL's magazines using Apache Solr. It's worth pausing here to explain the significance of an index. When using any search engine (including Google), you are searching an index, not the actual content.

A search index is a lot like the index at the back of a book—it stores information in a special way that helps you find what you are looking for quickly. Search indexes, like Solr, take a document and blow it apart into lots of small bits so that a computer can quickly search it. Solr comes loaded with the search algorithms that do most of the mathematical heavy lifting. What we still had to do, though, was define the fields and data types for our index and crosswalk data and metadata from the Internet Archive into Lantern's index.

After the index was in place, we began working on a visual interface that would make it easy for users to run queries. We made a mistake at this point by building the interface from scratch using just HTML, CSS, and PHP (compare Figure 16.2, the original interface, and Figure 16.3, the redesigned interface).

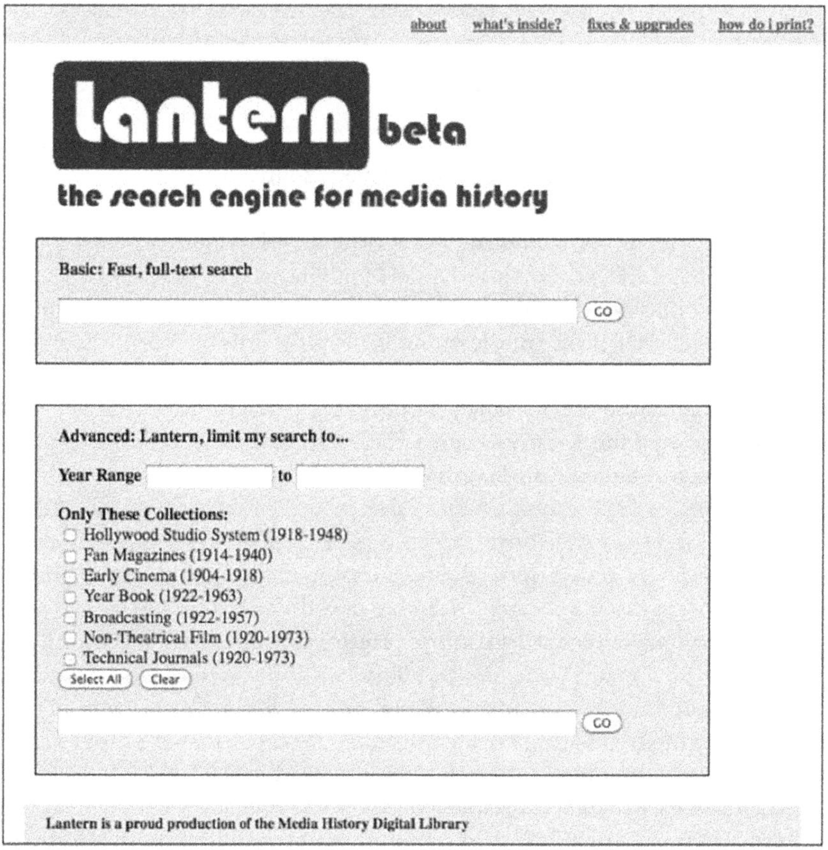

FIGURE 16.2 The prototype of Lantern's first interface, developed in 2011.

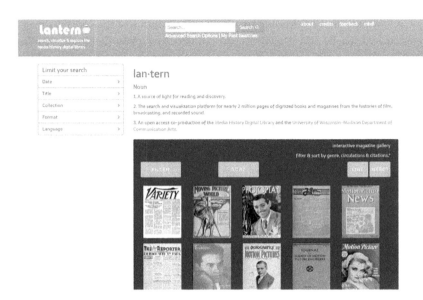

FIGURE 16.3   The redesigned version of Lantern, publicly launched in 2013. http://lantern.mediahist.org.

It was mistake because there were already open source interfaces for Solr that we could have used. We wasted a lot of time trying to recreate the wheel—only to eventually ditch our handmade rickety tire for Blacklight, an open source interface for Solr indexes developed by and for libraries. We were initially resistant to using Blacklight because we had zero experience with its programming language, Ruby on Rails. But learning Rails proved to be much easier than coding every tab and facet for an interface from scratch. Based on this experience, I would advise other scholars working on digital projects to adapt their workflows toward the best open source frameworks available, even if it requires learning a new programming language in the process.

In November 2011, I took Lantern's first prototype with me to Ann Arbor, Michigan for HASTAC (http://hastac.org), a conference I highly recommend for those interested in the digital humanities and applied media studies. I learned a lot at that conference, including how different our process had been from the way many digital humanities projects are formulated and led. I was surprised at how many academics and librarians I met viewed external grants as a pre-condition for working on a digital project. When I gave my brief presentation on the second day, I made the point, "You don't need a NEH digital humanities start-up grant to start up. You just need to start up." This was the line that audience members re-tweeted. And I still stand by that statement.

The following week, I took the Lantern prototype with me to Madison, Wisconsin to give a research presentation and interview for a faculty position in UW-Madison's Department of Communication Arts. Along with insights about the landscape for digital humanities projects, I had picked up something else at HASTAC: a bad cold. My voice was so gone by the time I reached Madison that I had to whisper into a microphone to make it through my job talk. I thought I had bombed and thrown away an amazing opportunity. But two weeks later, as I was finishing a regimen of antibiotics, I received a call from the department chair. Much to my surprise and delight, they offered me the job!

The MHDL and Lantern had taken up a great deal of my time that Fall. But the visibility of the MHDL and the technical skills I learned from working on Lantern also made me a stronger candidate for the job in Madison, which required someone who could succeed at a research-oriented university but also teach web design and video production. For academics, we always take a certain amount of risk when we spend time on a digital project rather than the more traditional path of conference papers, journal articles, and books. Sometimes the risk pays off, sometimes it doesn't. I'll always be grateful for the way it paid off for me at the moment in my life when I needed it the most.

Completing and Launching Lantern

When I moved to Wisconsin, Lantern's development moved with me. I continued to collaborate remotely with David, Carl, Wendy, Joseph Pomp, and Andy Myers (the latter two of whom generously volunteered their time to help with the project's technical side). But I found new collaborators, too—including an extremely supportive department chair, Michele Hilmes, and an extraordinarily talented and hard-working computer specialist, Peter Sengstock. I also benefited from start-up research funds from the University of Wisconsin-Madison's Graduate School that allowed me to purchase a web server for Lantern and hire three graduate students—Kit Hughes, Derek Long, and Tony Tran—to help add more material to the search index.

Despite obtaining university support, better hardware, and skillful collaborations, Lantern stalled for almost an entire year (May 2012–May 2013) before the pace of development picked up again in the summer of 2013. I was completely responsible for the lack of progress, in part for some understandable reasons (e.g., moving with a family to a new state, teaching brand new classes, trying to turn my dissertation into a book). But the lack of progress was also due to my own naiveté. I think I assumed that once I got the right people and technologies in place, the project would essentially finish itself.

To put things in baseball terms, I think I assumed I could simply be the general manager—the one who assembles the team and tries to point them in the right direction. But I could never get much accomplished as the GM. Even when I tried to be the manager—the leader who is there on the field, dressed in uniform, pinching runners and calling the bullpen—I still was not able to achieve my goals. Instead, I had to embrace the now extinct tradition of the player-manager, those position players who simultaneously took at-bats and gave the green lights to steal. In practice, this meant I had to spend a few hours every afternoon and evening that summer programming Lantern's index and interface, then use what I had learned and accomplished the next morning as a way to better direct my team of collaborators. During the periods when we were all actively working on the platform, something else positive happened— we all felt more motivated to keep pushing forward on the project. Unlike the player-manager who takes away a left fielder's spot in the line-up when he puts himself in the game, a project leader who dives into the technical minutiae only adds and does not subtract. And the help is usually appreciated by team members who are slogging through highly tedious and challenging technical tasks.

I can't say the extent to which my experience here is generalizable. There may be scholars who successfully build and sustain applied media studies projects while operating entirely in that GM or managerial role. And, judging by the clean organizational charts and streamlined work plans that I read about in many grant proposals, my process may be an aberration. What I can say, though, is that my project stalled when I wasn't on the field actively trying to solve problems, catch fly balls, and drive in runs. And I've seen other projects get stuck when their leaders are not actively working on addressing the project's biggest priorities and needs.

My slow pace in figuring these things out and making progress on Lantern resulted in one major benefit. If Wendy, Carl, and I had been successful in completing Lantern in the Fall of 2011, users would have only been able to search a roughly 100,000-page collection comprised primarily of *The Film Daily*, *Photoplay*, and *Business Screen*. By summer of 2013, however, the MHDL's collections had boomed in breadth, diversity, and size—up to 800,000 pages spanning 53 different publications. This growth happened because of David's skill in arranging mutually beneficial partnerships with institutions and organizations such as Library of Congress Packard Campus, Museum of Modern Art Library, Niles Essanay Silent Film Museum, Domitor, and others. And in Madison, Tony Tran and Derek Long further enriched the collection by spending weeks on the post-production and indexing of *Variety*, which the Wisconsin Historical Society scanned from microfilm on our behalf (high quality originals could not be obtained). The upshot was that a Lantern search in 2013 yielded exponentially more valuable results to the user than a Lantern prototype search in 2011. Equally important, the 2013 Lantern search returned results to the user in a fraction of the time that the prototype required, thanks to a Python algorithm coded by Edward Betts and Andy Myers.

At the end of July 2013, we officially launched Lantern. The public response was swift and very positive; we received nice write-ups on Indiewire and on renowned film scholar David Bordwell's blog, and many congratulatory social media comments. I wish I could say that I took time to appreciate the moment. But the truth is I immediately rushed into working on the book manuscript and journal articles that I felt woefully behind on. I was not sure whether the digital project that 34,858 people had used in its first ten weeks online would count toward tenure. One thing that helped, a few months later, was when our team received the Anne Friedberg Award for Innovative Scholarship from the Society for Cinema and Media Studies. The award had previously only gone to books; it gave a stamp of academic legitimacy to an unconventional project that was not published by a university press. The award also meant a lot to me, personally. When I had begun my Ph.D. studies at the University of Southern California, Anne Friedberg was the visionary and generous leader of my department. The Lantern team members who were able to attend the award presentation in Seattle took a group selfie, a wink to a stunt that Ellen DeGeneres had pulled just a couple weeks earlier when she hosted the Oscars (Figure 16.4).

If I could go back in time, I would let myself appreciate the initial summer launch much more, even with the uncertainty about tenure waiting in the wings. And I would also warn my thirty-year-old self: you've still got a lot more work ahead of you with this thing.

FIGURE 16.4 The Lantern team at the 2014 Society for Cinema and Media Studies Conference, Seattle, Washington. From left to right: Kit Hughes, Eric Hoyt, Tony Tran, Andy Myers, Joseph Pomp, Derek Long. Not pictured, but critical to the project's success were Carl Hagenmaier, Wendy Hagenmaier, Anne Helen Petersen, David Pierce, and Peter Sengstock.

Sustaining and Improving the Project

As I write this essay, in the fall of 2016, the work of sustaining Lantern has been going on for more than a year longer than it took to develop and launch the platform. And hopefully, many more years of keeping this project going will follow. The work of sustaining a digital project requires a patience and discipline to which most media studies scholars are unaccustomed. Humanities scholars generally stay hands off with the work of sustainability and preservation. And I get it. My first book, published in 2014, can sit on the library shelf and remain accessible to researchers for the next fifty years. But digital projects are different. Lantern, launched in 2013, requires regular maintenance and attention to remain accessible to users next year.

What does the labor of sustainability actually look like? The anecdote that opened this essay provides one example. Software and web hosting services break down from time to time. And when they do, I need to abruptly shift my priorities and try to find solutions immediately. Even when things are working fine, they could always be working better; Lantern contains some broken links and inaccurate metadata that users bring to my attention and I try my best to change. Most basically of all, when the MHDL scans more magazines, sustainability means that I need to index those magazines into Lantern so that they become searchable. This is a more complicated process than one might imagine due to how the MHDL's collections are organized on the Internet Archive. David Pierce, Derek Long, and other collaborators help with this indexing process, but I always complete the two final steps myself—in part because I use a specially configured computer in my office to complete them, in part because I have a hard time letting go.

One asset in sustaining and enhancing a digital project is the acquisition of grant funding. Lantern has not received any large federal or private foundation grants, but it benefited indirectly through Project Arclight (http:// projectarclight.org), a Digging into Data grant sponsored by the Institute of Museum and Library Services in the U.S. and Social Science and Humanities Research Council in Canada. Led by Charles Acland and myself from mid-2014 to mid-2016, Project Arclight developed a web-based app for exploring trends in media history through mining the MHDL's two million pages as a form of big data. For the Arclight app to be as useful as possible, we needed to search across more magazines, especially more years of *Variety*, than were then available in the MHDL. After adding those magazines to the Arclight framework, the extra step of indexing those files in Lantern was relatively quick and easy. A tip for project leaders looking for more direct sustainability funding: look at the NEH Office of Digital Humanities' new set of Digital Humanities Advancement grants, which can go toward "revitalizing and/or recovering existing digital projects" and encourage universities to commit matching funds and other resources toward these same goals (https://www.neh.gov/grants/

odh/digital-humanities-advancement-grants). The downsides to depending upon grants for sustaining digital projects are: (1) writing the grant proposals takes a lot of time; (2) generally, you wait several months after submitting them only to get rejected; (3) even if you are lucky and win, grants only cover a finite period of time and you will again be facing the challenge of sustainability after the funding ends.

Ultimately, I think the same claim I made about starting up all those years ago at HASTAC holds true for sustainability: you don't need a sustainability grant to keep things going, you just need to keep things going. Toward this end, I've found that sources of motivation can be just as important as sources of money. I have three main motivators that keep me actively engaged in Lantern's upkeep. The first is the other people I work with on the MHDL. When David or any of the MHDL's collaborators take steps that enhance the collection, I feel inspired to improve the collection's search portal. Second, I'm working on a book project that requires me to keep going back to Lantern for the sake of my own research. The book is a history of Hollywood's trade press, and I've been delighted by how many moments of serendipity have occurred—solving some technical problem that unexpectedly reveals a side of a publication that I had never seen before. Third, it's motivating to know that thousands of people use Lantern and depend upon it for their research (our Google Analytics show that between three and five hundred users visit per day, with the average user session lasting around eleven minutes). They are worth the time spent on updates and the headaches of tech problems.

FIGURE 16.5 The Media History Digital Library's new homepage, launched in March 2017. http://mediahistoryproject.org.

As I finish writing this essay, David and I are working on a major overhaul of the MHDL's websites—an initiative made possible thanks to the generous support of the Mary Pickford Foundation (Figure 16.5). We are also making more incremental improvements: adding newly scanned books and magazines; replacing thumbnail images that disappeared for several of the MHDL's most widely used journals. Rather than waiting for the next giant crash, it feels good to be proactive, spending time maintaining and improving something that I care a lot about. Come to think of it, that's a life lesson that extends beyond applied media studies.

Acknowledgements

This is a personal essay. But the extent to which the MHDL and Lantern are collaborative projects cannot be overstated. Thank you to David Pierce for founding the MHDL and inviting me to be part of it, and thank you to the MHDL's many sponsors and contributors for enabling the collection to expand to 2 million pages (and growing!). My appreciation extends to the team of collaborators that developed Lantern and have helped keep it going: Carl Hagenmaier, Wendy Hagenmaier, Kit Hughes, Derek Long, Andy Myers, Anne Helen Petersen, David Pierce, Joseph Pomp, Peter Sengstock, and Tony Tran. I would also like to thank the University of Wisconsin-Madison's Office of the Vice Chancellor for Research and Graduate Education and Department of Communication Arts for their support of my digital work.

Some Sections of This Essay Include Writing Reproduced and Reworked from Earlier Publications

Hoyt, Eric. (2013) "Bootstrapping a Digital Project? 5 Things to Consider," *The Spectator* 33(2): 31–37.

———. (2016) "Curating, Coding, Writing: Expanded Forms of Scholarly Production," in Charles R. Acland and Eric Hoyt (eds.), *The Arclight Guidebook to Media History and the Digital Humanities*, Falmer: REFRAME/ Project Arclight, 347–73. http://projectarclight.org/wp-content/uploads/ ArclightGuidebook.pdf

———. (2013) "Let's Talk About Search: Some Lessons from Building Lantern," *Antenna*, August 14.

Hoyt, Eric, Carl Hagenmaier, and Wendy Hagenmaier. (2013) "Media + History + Digital + Library: An Experiment in Synthesis," *Journal of Electronic Media Studies* 3(1). doi:10.1349/PS1.1938-6060.A.430

PART VII
Conclusion

17

CONCEPTUAL MODELS AND HELPFUL THINKERS

Kirsten Ostherr, Jason Farman, Anne Balsamo, Patrick Vonderau, Elizabeth Losh, Bo Reimer, Heidi Rae Cooley, Tara McPherson, and Eric Hoyt

Kirsten Ostherr

This chapter serves as a conclusion by way of annotated compilation, gathering the contributors' favorite resources for helping them, their colleagues, and their students do applied media studies work. I asked participants what conceptual models they have found helpful for extending and applying media theory as they move between making, writing, and teaching. Additional questions included, what articles, books, blogs, and Twitter streams would you urge readers to look into? And finally, what are the gaps in this field where you feel future research should be focused?

This chapter should be read as a provisional conclusion, in that this section, like the book as a whole, does not purport to be an exhaustive compilation of all publications that might be considered representative of "applied media studies work." The list of contributors to this volume includes many, but not all of the interesting scholar-makers whose work is redefining fields, or sub-fields, of a wide range of disciplines, including media studies, critical code studies, digital humanities, and others. Several recent and forthcoming manuscripts and collections focused on the intersections of media studies and digital humanities, such as Acland and Hoyt (2016), Gordon and Mihailidis (2016), Sayers (2018), and McPherson (2018) attest to the dynamism of this field at the present moment. It is my hope that this collection will provide another entry-point into the debates and experiments that are laying out new methodologies for hands-on, engaged research and can help future scholars move their experimental projects forward.

In addition to the scholars represented or already discussed in this collection, several important sites of applied media studies work around the globe are worth mentioning in particular: Marcel O'Gorman and Beth Coleman's (2008) work at the Critical Media Lab at University of Waterloo, Ontario, Canada; Matthew Fuller's (2007, 2012) projects at the Centre for Cultural Studies at

Goldsmiths College, University of London, England; Matt Ratto and Garnet Hertz's (2018) activities at the Critical Making Lab at the University of Toronto, Canada; the work of Anna Munster (2013) at the National Institute for Experimental Arts, University of New South Wales, Australia; and the collaborations at the MLab in the Humanities at the University of Victoria, Canada, led by Jentery Sayers (2018).

The contributions that follow provide resources for future reading and annotation of the authors' commentaries throughout the book, demonstrating the wide range of influences, from political theorists and philosophers to visual artists, sociologists, activists, citizen scientists, journalists and bloggers, that have shaped the applied media studies work represented in this collection. While there is some overlap between the participants' recommendations, there is also a surprising amount of divergence. This heterogeneity of scholarly framings plays an important role, I believe, in fostering the richness of the field of applied media studies at this historical moment, and it is my hope that the corpus of influences will continue to expand, not contract, as more scholars engage in hands-on, critical, and creative media practice.

Future entanglements in this field might focus on some of the gaps identified by our contributors, who noted the need for more work that analyzes and designs complex media systems; more creation of large-scale collaborative platforms; greater interconnections between U.S.-based and non-U.S. scholar-makers; increased emphasis on the role of historical and archival research in digital projects; and finally, more attention to the conditions of precarity among many scholar-makers in higher education who are working within contingent lines of funding and employment.

What Kinds of Conceptual Models Have You Found to Be Helpful As You Move between Making and Writing or Teaching About Making? What Are the Articles, Books, Blogs, and Twitter Streams That You Would Urge Readers to Look into?

Jason Farman

Conceptually, my work bridges the fields of critical making and iterative design. The field of critical making, drawing on scholarship by people such as Tara McPherson (2018), Garnet Hertz (2012), Jentery Sayers (2018), John Maeda (2017), Kim Sawchuk and Barbara Crow (2015), and Kari Kraus and Rachel Donahue (2012), among a whole host of others, approaches media objects as tools to think with. The process of creation is one that allows us to analyze issues such as race, class, gender, and sexuality through media objects. As I design these objects to think with, whether that be the underlined code making an app work or a 3D model that leads to a print of an obsolete technology, I employ an iterative design approach. This approach begins by working directly with people to understand how they use their technologies. By then moving through the prototyping stage to usability testing, the critical making that I engage in seeks to recognize that different bodies are addressed differently by these different technologies and media objects in the world. Often these realizations can be an end in and of themselves, as I am able to address the ways that these media objects create different modes of what I call "sensory-inscription;" that is, the ways that different bodies engage media in a sensory way while simultaneously being inscribed as particular bodies by these technologies.

After usability testing, drawing on my humanities background, I engage in a deep analysis of these tests and adjust my designs accordingly. This process is ongoing and each iteration reveals new relationships between particular bodies, particular cultures, and particular demographic relationships to the medium that I am creating. The methods of critical making and iterative design pair well with participatory design, in which the end users of a medium are involved with the creation of that medium from the very beginning. That is, the team that is designing the medium are the people who will ultimately use it on a daily basis. These methods are still often constrained by hierarchies within the academy and the various modes of privilege that inscribe themselves onto the bodies of those who are the primary investigators of these projects. Therefore, it's important to continually recognize the positions of the participants and their various levels of agency throughout the process. While the design processes I engage never assume a universal user, there are still many hurdles that these processes need to overcome in order to design for equity and inclusion.

Anne Balsamo

I make use of and take inspiration from the work of sociologists Susan Leigh Star and James Griesemer (1989), especially their notion of boundary objects, which they define as "information, such as specimens, field notes, and maps, used in different ways by different communities. Boundary objects are plastic, interpreted differently across communities but with enough immutable content to maintain integrity." I fold this concept (of the boundary object) into my interpretation of an articulation model of cultural formations (Gramsci 1973; Hall 1980). What results is a model of the "applied media studies project" as an assemblage of (disparate) elements that are articulated to one another through the agentic practices of humans and materials; some of the elements function as boundary objects that participate too in the negotiation of identities. See also Fuller (2007), Hayles (2002), and Suchman (2007).

Patrick Vonderau

I do not think there is a particular kind of conceptual model that is better suited than other types of models. It largely depends on the context and object of the research itself. In general, reading up on model theory (and on the philosophy of science) may be a useful starting point. In what ways do our conceptual, statistical, artistic, and other models speak to each other? What do they represent, how, generating which kind of knowledge, for which purpose? What do they leave out, and why?

Books like *Inventive Methods* (Lury and Wakeford 2012) or *Digital Research Confidential* (Hargittai and Sandvig 2015) are of great value to expand our understanding of methods and research frameworks. The same goes for the Digital Methods Initiative in Amsterdam (Rogers 2008, 2014). Nicole Starosielski's wonderful *Surfacing* project (2014, 2015) is one I frequently revisit.

Elizabeth Losh

The theoretical framework of intersectional feminism has been central to my work in this area. Although I live in Virginia now, my DH SoCal identity is still really important to me, so I would recommend looking at any of the writers at the Octavia E. Butler Legacy Project (2005) for ways to understand connections between social justice movements, science fiction, and Afrofuturist art; Curtis Marez (2016) on Chicano Futurism, and great SoCal films like Alex Rivera's *Sleep Dealer* (2008); the materials generated by Teddy Cruz (2013), Ricardo Dominguez (n.d.), and others in connection with Political Equator; and the work being done at the online journal *Catalyst: Feminism, Theory, Technoscience* (2015) at UCSD for feminist Science and Technology Studies (STS) research. Digital identity is important but so are the lived conditions of face-to-face material, embodied, affective, situated, and labor-intensive interpersonal interactions that take place in a particular regional context.

At the international level, the work of Sam Gregory (n.d.) at WITNESS is always worth reading for contemporary media ethics; Lisa Parks and Nicole Starosielski (2015) on global media infrastructure is also incredibly useful.

Bo Reimer

I am not sure that general conceptual models will work for all the complex environments that should be in focus for applied media researchers. But going back to ideas about learning by doing (Dewey 1916, 1927) and the reflective practitioner (Schön 1983) is always helpful. I find Bruno Latour's and Peter Weibel's *Making Things Public* (2005) inspirational, as well as Bruce Mau's "Massive Change" book and network (2005, 2016). I am also fond of the "In Media Res" (2006) project.

Heidi Rae Cooley

The "applied" work I do provides me an opportunity to pursue in more concrete ways the more abstract philosophical models to which I'm drawn. Currently, I'm interested in habit change as theorized by Charles Sanders Peirce (1934/1935). For Peirce, habit change is a semiotic process that is necessarily social; it transpires according to a community of interpreters or inquirers. If we have grown accustomed to our touchscreen devices locating and tracking us and, subsequently, pushing to us in real time, for example, recommendations based on our current location and patterns in our behavior (habits), I am interested in how those very same micro-processes might inspire new modes of thinking about and relating to place, to each other, and to the fact that we are constantly "on" and connected. The Ward One app, and *Ghosts* and *Augusta App* before it, allows me to consider how such habit change might be mobilized by means of the very technologies that function to produce and perpetuate (ensure even) routines that make us findable and our likely navigations (of content and place) determinable. Because I take seriously Peirce's argument that habit change involves a community of interpreters, the project is collaborative.

How I approach this "applied" work draws heavily on the studies-practice culture I experienced while a graduate student at the University of Southern California in the School of Cinema-Television (now the School of Cinematic Arts). My exposure to the Labyrinth Project, headed by Marsha Kinder (1997), and the Institute for Multimedia Literacy (IML), directed by Anne Balsamo, provided concrete examples of how the shape of humanistic scholarship was shifting. At the time, Balsamo was writing *Designing Culture: The Technological Imagination at Work* (2011), and the IML was piloting undergraduate curriculum in multimedia authorship. Meetings were also taking place with Bob Stein, whose company Night Kitchen had developed a multimedia authoring platform, TK3, which was further developed into Sophie at the Institute for the Future of the Book. MediaCommons (Fitzpatrick and Santos 2007/2009), a community network for scholars interested in alternative modes of publication, such as In Media Res and [In]Transition, is likewise a project of the Institute for the Future of the Book. Tara McPherson and Steve F. Anderson launched *Vectors: Journal of Culture and Technology in a Dynamic Vernacular* (first volume appearing in fall 2005), and I was able to participate in a summer workshop attended by contributors to the journal's third volume (fall 2007 and summer 2012). McPherson's and Anderson's work on *Vectors* (2005) coincided with the establishment of CriticalCommons (in 2008; Anderson 2012) and the journal served as a prototype for Scalar.

Beyond the projects mentioned above, applied media-oriented work that I have found to have historical and socio-cultural significance include the online sites *AIDS Quilt Touch* (Balsamo et al. 2012), *Mapping Ararat* (Shiff, Kaplan and Freeman 2016), *Edmonton Pipelines Project* (Zwicker et al. 2010), and "Go Queer," a geo-locative game in development that foregrounds the queer history of Edmonton in Alberta, Canada (Harley 2018).

Tara McPherson

Activist practice has been a strong inspiration for our lab. That includes a DIY spirit and a focus on practical interventions. It's also a model that allows participants to bring different skill sets to a shared space to work on concrete issues. Rather than building technology in abstraction, we are committed to building infrastructure for existing projects in order to address the needs and issues that progressive scholars highlight for us. I am also very interested in emerging models for arts-based research, particularly those that focus on collaboration across differences. Creating fair evaluation and support structures for such work is an important issue for the university today if we aim to grow and extend applied media studies in ethical ways.

A few key collaborative sites to point to include Hastac.org (Davidson and Goldberg 2002), FemTechNet (2012), and Jentery Sayers' Makers Lab (2018). Two authors I would emphasize are Anne Balsamo, especially *Designing Culture* (2008) and Miriam Posner (2009).

Eric Hoyt

I'll speak here to the building of digital collections—especially collections of digitized texts and media productions—as a particular area of applied media studies work. In a couple of recent essays, my Arclight collaborators and I have discussed Jerome McGann's book *A New Republic of Letters* (2014) as a useful model for understanding the scholarly contributions of digitization and digital collection building. McGann argues for the renewed importance of philology, which studies how texts and languages change over time. Philology's status within the humanities fell in the late nineteenth and early twentieth centuries as the work of creating critical editions of texts came to be perceived as less important than the work of interpreting texts. However, as McGann points out, philology's attention to how texts change across time and forms is valuable in our contemporary era of digitization. And because these practices are valuable, we need to value them as a field.

I've learned a ton from the other contributors to this book. Additional media studies scholars who I admire for their digital humanities work, public outreach, and prolific Twitter streams are Christine Becker @crsbecker; Cynthia Meyers @AnneHummert; Jason Mittell @jmittell; Anne Helen Petersen @annehelen; Miriam Posner @miriamkp; and Deb Verhoeven @bestqualitycrab. I would also encourage readers to look at the freely downloadable book that Charles R. Acland and I co-edited, *The Arclight Guidebook to Media History and the Digital Humanities* (2016), which features essays by Meyers, Posner, and Verhoeven.

What Are the Gaps in This Field Where You Feel Future Research Should Be Focused?

Anne Balsamo

We need more work on the analysis of media systems, and the design practices of creating complex media systems. Applied media studies projects are often conceptualized as object-based entities (an application, a website, an archive portal, an interactive essay, a digital edited collection). Media formations are complex assemblages that cohere in a specific way, determined in part by historically specific influences but also (in part) by design, chance, and serendipity. They are both "over-determined" and "under-determined." How might we understand applied media studies projects within the context of broader media formations?

Patrick Vonderau

I would like to see more large-scale collaborative platforms such as Zooniverse (Citizen Science Alliance 2009) in media studies. I would also like to see more projects that build bridges between "traditional" and "applied" work in our field, and between work done outside the U.S. and inside the U.S.

Elizabeth Losh

More archival work! We need to think about cultural memory when we talk about digital ephemera, and groups that study the history, rhetoric, or philosophy of technology are often too conservative to attend to these preservation and access questions.

Bo Reimer

There are many gaps; this is still an emerging field. Right now I would just like to see many different kinds of projects take off. And I would like to see many different attempts of trying to represent the work coming out of these projects in other ways than just the written text.

Heidi Rae Cooley

I am very interested to see how this volume affords a mapping of influences—sites of confluence and departure across approaches to or modes of applied media studies. How do scholar-practitioners, projects, initiatives, institutions, etc., articulate? In what ways do our accounts complement and/or counter each other? How do various histories—personal, project, institutional—become visible? How do these histories speak to the (kinds of) politics such work provokes, participates in, perpetuates, and/or suffers? In other words, I look forward to seeing how *Applied Media Studies* will contribute to revealing and opening gaps where new habits might take root.

Tara McPherson

I think the most pressing issues in the field are the issues shared by the larger academy. We need to be acutely aware that our experiments in applied media studies might dovetail with broader movements to corporatize the university.

Applied media studies can be a force to resist these trends, but there is no guarantee that practices of making will help democratize the university further. The trend toward corporate sponsorship of lab spaces should not be embraced without careful thought. We need to be attentive to increasing patterns of casualized labor, particularly since there is now a tendency to shunt "makers" into less stable forms of employment. We need to make sure that "making" is available outside the realm of R-1 universities at a variety of scales. In short, we need to ensure that our commitments to making line up with our commitments to social justice more broadly.

Eric Hoyt

I think there is an important conversation beginning to happen between media scholars and media archivists about the ways in which scholars can take a more active role in improving our understanding of historical works of media and our ability to access them. The Media Ecology Project, led by Mark Williams at Dartmouth College, is one important initiative in this regard. Mark and his collaborators are creating digital environments for access that scholars can use in the course of their research and, in the process, generate descriptive metadata that flows back to the archive. Jeffrey Jones and Ethan Thompson have also been working on expanding access to the Peabody Awards Archive at the University of Georgia, and in the process, challenging some long-held assumptions about television history. I'm excited to contribute to these growing initiatives, and I would love to see even more work happen in this space.

References

Acland, Charles R., and Eric Hoyt, eds. (2016) *The Arclight Guidebook to Media History and the Digital Humanities*, Falmer: REFRAME/Project Arclight. http://book. projectarclight.org.

Anderson, Steve. (2012) Critical Commons: For Fair & Critical Participation in Media Culture Website. www.criticalcommons.org/

Balsamo, Anne. (2011) *Designing Culture: The Technological Imagination at Work*, Durham, NC: Duke University Press.

Balsamo, Anne, Dale MacDonald, Nikki Dudley, Kayla Haar, Lauren Haldeman, Mark NeuCollins, Kelly Thompson, Jon Winet and the NAMES Project Foundation. (2012) *AIDS Quilt Touch*. http://aidsquilttouch.org/content/main-menu.

Catalyst: Feminism, Theory, Technoscience. (2015) Catalyst Lab, Department of Visual Arts, University of California, San Diego. http://catalystjournal.org/ojs/index.php/catalyst/index.

Citizen Science Alliance. (2009) Zooniverse Website. www.zooniverse.org/

Cruz, Teddy. (2013) "How Architectural Innovations Migrate across Borders," *TED Talk*. www.ted.com/talks/teddy_cruz_how_architectural_innovations_migrate_across_borders.

Davidson, Cathy, and David Theo Goldberg. (2002) HASTAC (Humanities, Arts, Science, and Technology Alliance and Collaboratory) Website. www.hastac.org/.

Dewey, John. (1916/1997) *Democracy and Education*, New York: The Free Press.

———. (1927/1991) *The Public and Its Problems*, Athens, OH: Swallow Press.

Dominguez, Ricardo. (n.d.) Thing.net Website. www.thing.net/~rdom/

FemTechNet. (2012) https://femtechnet.org/.

Fitzpatrick, Kathleen, and Avi Santos, eds. (2007/2009) MediaCommons. A Digital Scholarly Network. Institute for the Future of the Book, University of Southern California, Los Angeles, CA. http://mediacommons.futureofthebook.org/.

Fuller, Matthew. (2007) *Media Ecologies: Materialist Energies in Art and Technoculture*, Cambridge, MA: MIT Press.

Fuller, Matthew, and Andrew Goffey. (2012) *Evil Media*, Cambridge, MA: MIT Press.

Gordon, Eric, and Paul Mihailidis, eds. (2016) *Civic Media: Design, Technology, Practice*, Cambridge, MA: MIT Press.

Gramsci, Antonio. (1973) *Selections from the Prison Notebooks*, London: Lawrence & Wishart.

Hall, Stuart. (1980) "Encoding/Decoding," in Stuart Hall, Dorothy Hobson, Andrew Lowe, and Paul Willis (eds.), *Culture, Media, Language: Working Papers in Cultural Studies, 1972–79*. London: Hutchinson, 128–138.

Hargittai, Eszter, and Christian Sandvig, eds. (2015) *Digital Research Confidential: The Secrets of Studying Behavior Online*, Cambridge, MA: MIT Press.

Harley, Jason. (2018) "About the Lab: Current Research," CHI-TEA Laboratory. Computer-Human Interaction: Technology, Education, and Affect. http://chitealab.ualberta.ca/.

Hayles, N. Katherine. (2002) *Writing Machines*, Cambridge, MA: MIT Press.

Hertz, Garnet, ed. (2012) *Critical Making*, Hollywood, CA: Telharmonium Press. http://conceptlab.com/criticalmaking/.

In Media Res Project. (2006) Institute for the Future of the Book, University of Southern California, Los Angeles, CA. http://mediacommons.futureofthebook.org/imr/

Kinder, Marsha. (1997) "The Labyrinth Project," University of Southern California, School of Cinematic Arts, Los Angeles, CA. http://dornsife.usc.edu/labyrinth/.

Kraus, Kari, and Rachel Donahue. (2012). "'Do You Want to Save Your Progress?' The Role of Professional and Player Communities in Preserving Virtual Worlds," *Digital Humanities Quarterly* 6(2). www.digitalhumanities.org/dhq/vol/6/2/000129/000129.html.

Latour, Bruno, and Peter Weibel. (2005) *Making Things Public: Atmospheres of Democracy*, Cambridge, MA: MIT Press.

Lury, Celia, and Nina Wakeford, eds. (2012) *Inventive Methods: The Happening of the Social*, NY: Routledge.

Maeda, John. (2017) "Design in Tech Report." https://designintechreport.wordpress.com/.

Marez, Curtis. (2016) *Farm Worker Futurism: Speculative Technologies of Resistance*, Minneapolis, MN: University of Minnesota Press.

Mau, Bruce. (2016) *Bruce Mau's Massive Change Network Website*, Evanston, IL. www.massivechangenetwork.com/.

———. (2005) *Massive Change*, London: Phaidon Press.

McGann, Jerome. (2014) *A New Republic of Letters: Memory and Scholarship in the Age of Digital Reproduction*, Cambridge, MA: Harvard University Press.

McPherson, Tara. (2018) *Feminist in a Software Lab: Difference + Design*, Cambridge, MA: Harvard University Press.

McPherson, Tara, and Steve Anderson, eds. (2005) *Vectors: Journal of Culture and Technology in a Dynamic Vernacular*, Los Angeles, CA: University of Southern California. http://vectors.usc.edu/journal/index.php?page=EditorialStatement.

Munster, Anna. (2013) *An Aesthesia of Networks: Conjunctive Experience in Art and Technology*, Cambridge, MA: MIT Press.

O'Gorman, Marcel, and Beth Coleman. (2008) Critical Media Lab, University of Waterloo, Ontario, Canada. http://criticalmedia.uwaterloo.ca/crimelab/.

Octavia E. Butler Legacy Network. (2005) http://octaviabutlerlegacy.com/.

Parks, Lisa, and Nicole Starosielski, eds. (2015) *Signal Traffic: Critical Studies of Media Infrastructures*, Chicago: University of Illinois Press.

Peirce, Charles Sanders. (1974) [1934/1935] *Collected Papers*, vol. V and VI, Charles Hartshorne and Paul Weiss (eds.), Cambridge, MA: The Belknap Press of Harvard University Press.

Posner, Miriam. (2009) Miriam Posner's Blog Website. http://miriamposner.com/blog/.

Ratto, Matt, and Garnet Hertz. (2018) "Critical Making and Institutional Learning: Making as a Bridge between Art, Science, Engineering, and Social Intervention," *Leonardo Electronic Almanac*. www.leonardo.info/leonardo-electronic-almanac.

Rogers, Richard. (2013) *Digital Methods*, Cambridge, MA: MIT Press.

————. (2008) Digital Methods Initiative, University of Amsterdam, Netherlands. https://digitalmethods.net/

Sawchuk, Kim, and Barbara Crow. (2015) "Ageing Mobile Media," in Gerard Goggin and Larissa Hjorth (eds.), *Routledge Companion to Mobile Media*, New York: Routledge, 249–76.

Sayers, Jentery, ed. (2018) *The Routledge Companion to Media Studies and Digital Humanities*, New York: Routledge.

Schön, Donald A. (1983) *The Reflective Practitioner: How Professionals Think in Action*, New York: Basic Books.

Shiff, Melissa, Louis Kaplan, and John Craig Freeman. (2016) *Mapping Ararat. An Imaginary Jewish Homelands Project*. www.mappingararat.com/.

Star, Susan, and James Griesemer. (1989). "Institutional Ecology, 'Translations' and Boundary Objects: Amateurs and Professionals in Berkeley's Museum of Vertebrate Zoology, 1907–39," *Social Studies of Science* 19(3): 387–420.

Starosielski, Nicole. (2015) *The Undersea Network*, Durham, NC: Duke University Press.

Starosielski, Nicole, Eric Loyer, and Shane Brennan. (2014) *Surfacing*. www.surfacing.in

Suchman, Lucy. (2007) *Human-Machine Reconfigurations: Plans and Situated Actions*. New York: Cambridge University Press.

Williams, Mark. (2013) Media Ecology Project, Dartmouth College. http://sites.dartmouth.edu/mediaecology/introduction-to-the-media-ecology-project/.

WITNESS website (n.d.) https://witness.org/.

Zwicker, Heather, Daniel Laforest, Maureen Engel, and Russell Cobb. (2010) *Edmonton Pipelines Project. Maps and Narratives for Dense Urban Spaces*. University of Alberta, Canada. http://edmontonpipelines.org/.

INDEX

NOTE: references to figures appear in *italics*

Academia.edu 132
accountability 34, 203–216, 213
Acland, Charles 248
activism 154, 155
Agamben, Georgio 123
agonism 203–216
Ahmed, Sara 90
Ahn, Mollie *71*
AIDS Memorial Quilt *142 see also AIDS Quilt Touch (AQT)*; databases 144, 145; first panel of *144*; searching *145*
AIDS Quilt Touch (AQT) 182, 187, 222; AIDS Quilt Touch Interactive Timeline *154*; AIDS Quilt Touch Virtual Quilt Tabletop: Browser *149, 150, 151*; background of 142–156; collaborations across differences 141–158; launch *152*; lessons from 156–157; McMullin, Gert *147*; phases of project 146; Quilt Index 155; searching *145*; Virtual Quilt images *148*
AIDS Quilt Touch Indigogo campaign *156*
AIDS Quilt Touch Interactive Timeline *154*
AIDS Quilt Touch Virtual Quilt Tabletop; Browser *149, 150, 151*
Airtel 111
Algorithmic Design II (CSCE 146) 138
algorithms 13, 18, 134, 195, 197, 227, 246
Alliance for Networking Visual Culture 55
Alvarado, Rafael 23
American Quarterly 7
American Studies Association 7
Anderson, Steve 49, 53, 54, 57, 85

Angewandte Medienwissenschaften 35
animation 28, 36, 63, 85
Anne Friedberg Award for Innovative Scholarship 247
Apache Solr 242, 243
Apple 203
application programming interfaces (APIs) 52
applied humanities 6, 7, 10, 19, 237
"Applied Humanities" (Steinberg) 6
"The Applied Humanities: A Businesslike: Approach" (Kolbert) 7
applied sciences 3, 4–9
apps 52, 90, 137, 237; *AIDS Quilt Touch (AQT) see AIDS Quilt Touch (AQT)*; Ward One *see* Ward One
Arclight 236
Atlas of Emotion (Bruno) 112
audio documentary *(Public Secrets)* 134
authoring tools 58 *see also* Scalar
autodidacticism 13

backgrounds, disciplinary 223–228
Bailey, Moya 183
Balsamo, Anne 17, 38, 69, 118; advice 237; attaining resources 222; career advancement 232; change in roles 190; conceptual models 255; education, mentoring, training 227; ethics (unintended consequences) 182; gaps in field (of applied media) 258–259; meaning of applied media studies 35; online ethical challenges 187; principal

investigators 229–230; reimagining
humanistic media studies 43
Baraniuk, Rich 69
Barney, Darin 98
Baym, Nancy 196
Betts, Edward 246
Big Data 88, 193, 195
Björgvinsson, Erling 206
Blacklight 244
Black Lives Matter movement 183
Blaine, Keisha N. 120
Blake, William 25
Boal, August 47
Bolivar, Bill 153
Bolivar, Mark 153
Bolter, Jay David 47
Book Riot 122
Bordwell, David 246
Bowker, Geoffrey 38
boyd, danah 196
Brakhage, Stan 4
Brandt, Lydia 169
#BrexitSyllabus 121
Brown, Michael 119
Brown University's English Department 25
Bruno, Giuliana 112
Bud, Robert 5
Buell, Duncan 90, 133, 160, 161, 168, 184
Bully Bloggers 120
Burgess, Henry 196
Burrell, Jenna 97
Busa, Roberto 24
Business Screen 246
Bustamante, Carlos 36

California Youth Authority 37
career advancement 230–233
Casey, Travis *175*
Caughman, Richard 175
Center for Digital Humanities 96, 133
Center for Solutions to Online Violence 184
#Charlestonsyllabus 120, 121
Chatelain, Marcia 119, 124
Chief Macha, interview of 103, 104, 105
Chow, Mark 144
Christen, Kimberly 53
Christian, Barbara 123
Chun, Wendy 55
classrooms, flipped model 73, 74
Clutterbuck-Cook, Hannah 122
Coalition for Networked Information 55
co-authoring 131 *see also* collaborations
code for America 3
Coleman, Beth 83, 253

Coleridge, Samuel Taylor 5
collaborations 3, 17, 32, 34; across
differences 141–158; *AIDS Quilt Touch
(AQT)* 141–158; challenges of 84;
change in roles 190–191; Connected
Courses 132; cross-college 159–178;
cultivating communities of interpreters
161–169; design 51; expanding
communities of interpreters 169–171;
faculty 169; favorite part of applied
media studies 82–88; FemTechNet
83, 132; hands-on collaborative work
81–96; impact of 88–91; information
and communication technology
(ICT) research 97–116; landscapes 42,
43; Macha, Zambia 101–102; media
interventions 205–211; Medical Futures
Lab 61, 66 *see also* Medical Futures
Lab; online ethical challenges 187–189;
process of 131; rapid response 117;
Scalar 58; Selfie Researchers Network
132; stakeholders 87; technology
51; time, energy, effort of 91–94;
transdisciplinary 129–140; unintended
consequences 181–191; *Vectors* 49–51;
with Ward One members 171–176
collaborators, finding 130–133
College Art Association 56, 92
communication 3, 4, 5, 9, 16, 32, 35, 36, 37,
38, 52, 53, 67, 70, 71, 90, 94, 205–211;
difficulties 82; reconfiguration of 35;
sniffing out *199*; with users 87
community engagement 101–102
*Companion to Media Studies and Digital
Humanities* (Sayers) 14
computer scientist collaborations 32
concept mapping *163*
conceptual models 253–262
concordances 24, 25
Connected Courses, collaborations 132
content, brainstorming *163*
The Conversation (theconversation.com) 138
Cooley, Heidi Rae 17; advice 234;
attaining resources 220–221; career
advancement 230–231; change in roles
190–191; conceptual models 256–257;
education, mentoring, training 224–225;
ethics (unintended consequences)
184–185; favorite part of applied
media studies 87; finding collaborators
133; gaps in field (of applied media)
259; impact of collaboration 90–91;
institutional settings 96; meaning of
applied media studies 32–33, 42; online

ethical challenges 188–189; principal investigators 228–229; time, energy, effort (of collaboration) 93; translation between fields 138–140
Cooper, Mark 169
copyright protection 40, 185, 240
Corbin, Betty 175
core objectives, Medical Futures Lab 64
core values, Medical Futures Lab 64
Couldry, Nick 98
Course Tumblr page for "Medicine in the Age of Networked Intelligence" 68
Cowan, Ruth Schwartz 6
Cowan, T.L. 229
Craib, Raymond B. 46
Critical Code Studies 13 critical contexts 10–14
Critical Interactives 17, 97, 140; Cooley, Heidi Rae as teacher 90, 93; support of 139, 161, 168, 170, 171, 221
Critical Making (Hertz) 12
critical thinking 68–71
cross-college collaboration 159–178; collaborating with Ward One members 171–176; cultivating communities of interpreters 161–169; expanding communities of interpreters 169–171
Crossing Boundaries (Klein) 51
Crow, Barbara 254
cultivating communities of interpreters 161–169

Daniel, Sharon 134, 171
Dartmouth College 260
databases, AIDS Memorial Quilt 144, 145
Debates in the Digital Humanities (Gold and Klein) 27
demonstrations, end-of-term *168*
Derrida, Jacques 7
design 9, 10, 11, 17, 19, 33, 36, 37, 43, 48, 49, 50, 51, 68–71, 87, 96, 148, 182, 187, 222; collaborations 49, 51; meetings 50; tensions 51
Design Cultures & Creativity 95
development: Critical Interactives 171; Lantern 238–250; processes 134; of shared languages 159–178
Dewey, John 73, 205
diagrams: groups *164, 165*; user experience: (UX) *165, 166*
Dietrich, Craig 49, 53, 54, 85, 139
difference, collaborations and 49–51
Digital Dynamics Across Cultures for Vectors: (Christen) 189

digital humanities (DH)11, 13, 14, 23–28, 36, 89, 193, 222; definition of 23; differences between applied media and 36; investigations 27; methods 24; role of 23
Digital Humanities Grants (NEH) 222
Digital Humanities Quarterly 23 digital intermediation 3–22, 5
Digital Library Federation 120
digital medical humanities 60–78; advice 76–77; Connected, Online, Hands-On- the Digital *versus* the Analogue 72–74; Creative, Hands-on, Critical Thinking and Design 68–71; Imagining the Medical Futures Lab 64–66; Medical Futures Lab in Historical Context 74–76; Project TMC (Medical Futures Lab) 61–63; Tooling Up 66–68
digital music distribution 198
Digital Research Confidential (Hargittai and Sandvig) 256
digital scholarship 51, 55
digitization 5, 98, 186
disciplinary backgrounds 223–228
Distributed Open Collaborative Course (DOCC) 38, 74, 118
DIY (do-it-yourself) curricula 117–124
do-it-yourself (DIY) 3, 12, 13, 17
Donahue, Rachel 254
Donaldson, Bobby 139, 170, 172
Dourish, Paul 49
Dublin Institute of Technology 192

Earhart, Amy 183
Eaves, Morris 25
education 223–228
Egan, Carolee 153
electronic health record (EHR) systems 60, 76
electronic text encoding (ETE) 25
end-of-term presentations *168*
The Equality Lab 222
Eriksson, Maria 196, 199
Essick, Robert 25
ethics18, 62, 66, 73, 81, 182, 195, 196, 211–214, 256; new research media landscape 211–214; online ethical challenges 187–189; unintended consequences 182–187
Ethington, Phil 49
European Union (EU) 122
exercises, programming 167
expanding communities of interpreters 169–171

experimentation: 4, 51–53, 55, 72, 86, 94, 134, 162, 224, 235; digital medical humanities 60–78 *see also* digital medical humanities; and process 51–53
exploitation, conditions of 39
external relationships, cultivating 54–56

Facebook 56, 65, 75, 200, 203; groups 118
faculty: collaborations 169; metrics for evaluation 92
Farman, Jason 14; advice 235–236; career advancement 231–232; conceptual models 254–255; education, mentoring, training 226; ethics (unintended consequences) 182–183; favorite part of applied media studies 85–86; finding collaborators 131–132; impact of collaboration 90; institutional settings 95–96; meaning of applied media studies 41–42; reimagining humanistic media studies 46–47; time, energy, effort (of collaboration) 93; translation between fields 136–137
Feibleman, James 6
feminist media studies 45, 46
Feminist Science and Technology Studies 10
FemTechNet 38, 39, 74, 83, 94, 188; collaborations 83, 132; DIY; (do-it-yourself) curricula 117–124
fields, translation between 133–140
fieldwork 10, 16, 44, 82, 91, 92; baggage of history and 99; media 97–116
"Fight the Blight" campaign 171
film 4, 1124, 27, 32, 35, 36, 40, 41, 49, 56, 58, 63, 67, 75, 85, 96, 110, 123, 135, 139, 169, 208, 211, 239; theory 41
Film and Media Theory (FILM 473) 139
Film Daily 239, 242, 246
filming interviews *174, 175*
Films that Work: Industrial Film and the Productivity of Media (Hediger and Vonderau) 11
financial resources 219, 228–230
Finding Augusta: Habits of Mobility and Governance in the Digital Era (Cooley) 139, 230
Fitzpatrick, Kathleen 25
Flash 37, 57, 85
Fleischer, Rasmus 199
Fletcher, Curtis 48, 49, 54, 186
flipped classroom model 73, 74
Folk Life Festival (Smithsonian Museum) 148
Fossati, Giovanna 37

foundations (applied media) 31–47; Balsamo, Anne 35, 43; Cooley, Heidi; Rae 32–33, 42; Farman, Jason 41–42, 46–47; Hoyt, Eric 39–40, 44–45; Losh, Elizabeth 37–39, 44; McPherson, Tara 40–41, 45–46; meaning of 32–42; Ostherr, Kirsten 31; reimagining humanistic media studies 42–47; Reimer, Bo 33–35, 42–43; Vonderau, Patrick 35–37, 43–44
frameworks: for design meetings 50; Medical Futures Lab 66
Friedberg, Anne 247
Fuller, Matthew 254
funding 95, 238, 239

Galison, Peter 43, 44, 138, 193, 194, 196
Gaming the Humanities 160
gaps in field (of applied media) 258–260
gay rights 142, 143
Geertz, Clifford 225
Ghana 97
Ghosts of the Horseshoe (Ghosts) 161, *165, 170, 185, 190, 221*
Gillespie, Tarleton 196
globalization 98, 99
Gold, Matthew 27
Goldfarb, Brian 55
Goldsmiths College 254
Golumbia, David 54
Google 200, 203; Maps 57; Scholar 132
Graham, Lindsay 16
Gramsci, Antonio 42, 46
Griesmer, James 255
Grinberg, David 16, 17
groups: diagrams *164, 165*; Facebook 118; group work *164*
Gruner, Derek 170
Grusin, Richard 47

hackathons 3
Hagenmaier, Carl 242
Hagenmaier, Wendy 240
Hall, Gary 54
Handheld Art 191
hands-on collaborative work 81–96; favorite part of applied media studies 82–88; impact of collaboration 88–91; time, energy, effort (of collaboration) 91–94
hands-on critical thinking 68–71
hands-on productive work (making) 12
hands-on projects 3
Harris-Lowe, Bonnie *175*
Harvard 37, 118
hashtag activism 137, 138

hashtag syllabi 117–124
HASTAC 244, 249
Hayles, Katherine 27
healthcare reform 60 *see also* digital
  medical humanities
Health Information Technology for
  Economic and Clinical Health
  (HITECH) 60
Heidegger, Martin 46
Hertz, Garnet 12, 254
Hillgren, Per-Anders 206
historians, collaborations 32
historical context, Medical Futures Lab
  74–76
Historic Columbia Foundation 139
history: of activism 154; of AIDS 155
HIV-AIDS pandemic 143, 187 *see also*:
  AIDS Quilt Touch (AQT)
Holt, Jennifer 37
hospitals *104*
Hoyt, Eric 18; advice 236–237; conceptual
  models 258; education, mentoring,
  training 227; ethics (unintended
  consequences) 185–186; favorite part
  of applied media studies 87–88; finding
  collaborators 130; gaps in field (of
  applied media) 260; meaning of applied
  media studies 39–40; online ethical
  challenges 189; reimagining humanistic
  media studies 44–45; time, energy, effort
  (of collaboration) 93–94; translation
  between fields 133–134
HTML (hypertext markup language) 85
Hughes, Kit 245
human computer interaction (HCI)
  136, 151
Human-Computer Interaction Lab
  (HCIL) 132
humanistic media studies 3
humanistic research models 81
humanities 4–9
Humanities Gaming Institute 133
Huybrechts, Liesbeth 212

ICDL (International Computer Driver's
  License) 103
ICT4D researchers 97, 98
Image and Logic (Galison) 194
images 36, 38, 45, 50, 63, 74, 75, 87, 91,
  143, 146, 149, 151, 153, 155, 159, 189,
  250; Virtual Quilt *148*
impact of collaboration 88–91; Cooley,
  Heidi Rae 90–91; Farman, Jason 90;
  Losh, Elizabeth 88–89; McPherson, Tara

89–90; Parks, Lisa 88; Reimer, Bo 89;
  Vonderau, Patrick 88
Index Thomisticus (Busa) 24
information and communication
  technology (ICT) 69, 97–116
information technology (IT) 198
  infrastructural inversion 38; innovation
  9, 29, 56, 62, 63, 66, 67, 73, 129, 207,
  212; role of serendipity in 62
Instagram 75
The Institute for Advanced Technology in
  the Humanities (IATH) 26
Institute for Multimedia Literacy (IML)
  89, 220, 257
institutional settings: Cooley, Heidi
  Rae 96; Farman, Jason 95–96; Losh,
  Elizabeth 94–95; McPherson, Tara
  95; Reimer, Bo 95; role of 81, 94–96;
  Vonderau, Patrick 94
intellectual property issues 182–187
interfaces: application programming
  interfaces (APIs) 52; Lantern *243, 244*;
  Media History Digital Library (MHDL)
  249; user interface (UI) 162, 167, 185
intermediation, digital 3, 5, 60
internet, interviews via 102
Internet Archive 242, 243
The Internet of Things 203
interpreters: cultivating communities of
  161–169; expanding communities of
  169–171
interventions 17, 34, 82, 83; applied
  sciences 4–9; collaborative media
  205–211; critical contexts 10–14; for
  the digitally intermediated age 3–22;
  humanities 4–9; media 203–216;
  methods 14–15; technological 4
interviews: by community researchers
  106–110; Jones, Arthur *174, 175*;
  Macha, Zambia 102–103, *107, 108*;
  media fieldwork 102–103; by visiting
  researchers 103–106
Inventive Methods (Lury and Wakeford) 256
Irish, Sharon 229

Jenkins, Henry 8, 196
Jones, Arthur *174, 175, 175*
Journal of Popular Culture 7
Juhasz, Alexandra 38, 118

Kant, Immanuel 5
Kay, Alan 36
Kelly, Raegan 49, 52, 54, 133
Khan Academy 74, 75

Killoran, Peter 67, 68
Kinder, Marsha 49, 257
Kirschenbaum, Matthew G. 24
Klein, Julie 51
Klein, Lauren 27
The Knowledge Foundation 223
knowledge production 35
Kraus, Kari 254
Kurin, Richard 141

laboratory knowledge 43
Labyrinth Project 49
Lam, Alex *71*
Lantern 18, 45, 87, 88, 93, 134, 236;
    addressing a need 239–242; completing
    and launching 245–247; development
    238–250; interfaces *243, 244*;
    prototypes 244; reason for development
    of 242; starting up 242–245; sustaining
    and improving 248–250; team *247*
Larkin, Brian 97
Latour, Bruno 204
launches: *AIDS Quilt Touch (AQT) 152*;
    Lantern 245–247
LCD screens 33
Lead Technical Designers 130
Lee, Muriel 176
legal issues (unintended consequences)
    182–187
lesbian rights 142, 143
Les maitres fous (film) 110
LGBTIQVirginians (the Mattachine
    Project) 222
L'Hommedieu, Andrea 170
Lim, Eng-Beng 120, 124
LinkedIn 10
LinkNet 100, 101
Linknet Information Technology Academy
    (LITA) 101, 106
liquid net publishing 193
Liu, Alan 27
LiveLeak.com 52
live streaming 138
Living Labs 136
Long, Derek 245
Los Angeles Times 141
Los Angeles Unified School District 56
Losh, Elizabeth 17; advice 236; attaining
    resources 222; career advancement
    232; change in roles 190; conceptual
    models 256; education, mentoring,
    training 226–227; ethics (unintended
    consequences) 183–184; favorite
    part of applied media studies 83–84;

finding collaborators 132; gaps in
    field (of applied media) 259; impact
    of collaboration 88–89; institutional
    settings 94–95; meaning of applied
    media studies 37–39; online ethical
    challenges 187–188; principal
    investigators 229; reimagining
    humanistic media studies 44; time,
    energy, effort (of collaboration) 91–92;
    translation between fields 137–138
Loyer, Eric 49, 54, 85, 130, 134

MacArthur Foundation 220
MacDonald, Dale 144
Macha, Zambia: collaborations 101–102;
    community engagement 101–102;
    dirt road *100*; hospital *104*; interviews
    102–103, *107, 108*; interviews by
    community researchers 106–110;
    interviews by visiting researchers
    103–106; Macha Central School *105*;
    Macha Girls School *107, 111*; media
    fieldwork in 99–113; mobile phone
    towers *112*; site visits and path walking
    110–113
Macha Works 100
Macromedia 37
Madden, Ed 191
Maeda, John 9, 254
maker movements 3, 12
making (hands-on productive work) 12
Malmö City Symphony 208–214, *209,
    210, 211*
Malmö University 131, 205–211, 223
Mamber, Steve 36
Mann-Simmons House 174
mapping *163*
Markovitz, Lauren 36
Marwick, Alice 118
Maryland Institute for Technology in the
    Humanities (MITH) 132
Master of Public Health (MPH) degree 67
McCarthy, Anna 98
McGann, Jerome 26, 258
McMullin, Gert 146, *147*
McPherson, Tara 14, 16; advice 233;
    attaining resources 220; career
    advancement 230; change in roles
    191; conceptual models 257–258;
    education, mentoring, training
    223–224; ethics (unintended
    consequences) 186–187; favorite part
    of applied media studies 84; finding
    collaborators 130–131; gaps in field

(of applied media) 259–260; impact of collaboration 89–90; institutional settings 95; meaning of applied media studies 40–41; online ethical challenges 189; principal investigators 228; reimagining humanistic media studies 45–46; time, energy, effort (of collaboration) 92–93; translation between fields 134–135
meaning of applied media studies: Balsamo, Anne 35; Cooley, Heidi Rae 32–33; Farman, Jason 41–42; Hoyt, Eric 39–40; Losh, Elizabeth 37–39; Reimer, Bo 33–35; Vonderau, Patrick 35–37
Medea Talks 135
Medea Vox 135
media 239; ecosystems 10; interventions 203–216; new research landscape 203–204; production 10; studies 11
media archeology 98
Media Ecology Project 260
media fieldwork 97–116; community engagement 101–102; interviews 102–103; interviews by community researchers 106–110; interviews by visiting researchers 103–106; in Macha, Zambia 99–113; site visits and path walking 110–113
Media History Digital Library (MHDL) 39, 45, 134, 185, 189, 236, 239, *249 see also* Lantern
media researchers: roles of 214–215
Media Rituals (Couldry) 98
MediaSpace: Place, Space, and Culture in a Media Age (Couldry and McCarthy) 98
Medical Futures Lab 60–78; advice 76–77; collaborations 66; competencies 66; Connected, Online, Hands-On-the Digital *versus* Analogue 72–74; Creative, Hands-on, Critical Thinking and Design 68–71; development of 60, 61; frameworks 66; Imagining the Medical Futures Lab 64–66; Medical Futures Lab in Historical Context 74–76; Project TMC 61–63; social media 65; Tooling Up 66–68
medical settings 205–206
"Medicine in the Digital Age" MOOC poster *73*
meetings: design 50; Ward One *173*
Mellon Foundation xiii, 10, 55, 220

mentoring 18, 69, 94, 133, 160, 219, 223–228, 233
metadata 148, 198
Metaphysische Anfangsgrunde der Naturwissenschaft (Metaphysical Foundations of Natural Science [Kant]) 5
methods 3, 14, 32; digital humanities (DH) 24; interventions 14–15; inventing 82
metrics for faculty evaluation 92
Michaels, Eric 98
Microsoft Research 156
Middle East 99
Miller, Alexa 69
Miller, Lucas 49
Minneman, Scott 144
Mirzoeff, Nicholas 55
Mobile Interface Theory (Farman) 86
mobile phones: number discarded daily 183; repair costs 182; towers *112*; Zambia 105
mobile platforms 10
The Mobile Story: Narrative Practices with Locative Technologies 232
models: conceptual 253–262; flipped classroom 73, 74; humanistic research 81; teaching 81, 93
Modern Language Association (MLA) 53
MOOCs (Massive Open Online Courses) 38, 61, 72, 74, 118, 119
Mouffe, Chantal 212
Moving Image Research Collections and Digital Collections 96
Moving Picture World 239
Mudenda, Consider 101, 102, 103, *106,* 109
Mukurtu 53
Mulvey, Laura 41
Munster, Anna 254
music 199, 200
Myers, Andy 246

Nader, Laura 194, 195
Nakamura, Lisa 229
NAMES Project Foundation *145,* 148, 182
National Endowment for the Humanities: (NEH) 10, 55, 145, 155, 222, 248
National Institute for Experimental Arts 254
National Library of Sweden 197
National Science Foundation (NSF) 10, 221
Negroponte, Nicholas 36
A New Republic of Letters (McGann) 258

networks 5, 10, 12, 15, 38, 39, 56, 58, 104, 112, 118, 119, 184; efficiency 101
New Public Management 234
new research media landscape 203–204; See the Future project 206–208; Malmö City Symphony 208–211; Malmö University 205–211; politics and ethics 211–214
Nickelodeon Theater 139
Nigeria 97
Nwabueze, Obi *71*

objectives, Medical Futures Lab 64
Octavia E. Butler Legacy Project 256
O'Gorman, Marcel 253
on-demand culture 197
online ethical challenges 187–189
Open Access publishing 133
open educational resource (OER) 118
Open Source libraries 185
open syllabus paradigm 118
#OrlandoSyllabus 120
Ostherr, Kirsten: applied media 31; design problems 69; development of Medical Futures Lab 16; essays 16; as moderator 14; product engagement 11; scientific practices of healing 63; transdisciplinary collaborations 129–130
outcomes: imagining broader 82; unexpected 58
Oxford University Press 25

Paik, Nam June 4
Palmer, Lindsay 16, 17
Panfilov's 28 193
Parascope 207, *207,* 208
Parks, Lisa 14, 16, 38; advice 235; attaining resources 221–222; career advancement 231; education, mentoring, training 225–226; favorite part of applied media studies 82; finding collaborators 131; impact of collaboration 88; principal investigators 229; time, energy, effort (of collaboration) 91; translation between fields 136
Patheja, Jasmeen 190
path walking 110–113
PEN (the Public Electronic Network) 37
personal computers, rise of 24
Photoplay 240, *241,* 246
Pierce, Charles Sander 191, 256
Pierce, David 185, 240, 246, 249, 250 *see also* Media History Digital Library (MHDL)
pilot projects 63

platforms 195, 197, 242; archiving 155; authoring 85, 135, 257; collaborations 65, 254; for collecting behavioral user data 197; commercial 56; design of 10; edX 72; ethical responsibilities of 186, 195; to identify partners for dialogue 133; interoperability 38; launching 248; leveraging global 74; mixture of digital 4; monitoring 186; multi-platform access 153; open publishing 59; social media 72; streaming 199; teleconferencing 38; Wordpress-based *Book Riot* 122
podcasts 135
#PokemonGoSyllabus 121
politics, new research media landscape 211–214
Posner, Miriam 58
precarity, conditions of 39
presentations, end-of-term *168*
principal investigators 219, 220, 228–230
process, experimentation and 51–53
Profession (Anderson and McPherson) 53
programming exercises 167
Project Arclight 248
promotions 230–233
prototypes, Lantern 244
public interactives 148 *see also AIDS Quilt Touch (AQT)*
Public Secrets (audio documentary) 134
#PulseOrlandoSyllabus 120
pure science 6 *see also* science pure *versus* applied 7
Python 246

qualitative analysis of small data 193
qualitative social research 194
quantification 5
Queer Phenomenology (Ahmed) 90
Quilt Index 155
Quilt in the Capital event 153

racial inequality 171
rapid response 117–124
Ratto, Matt 254
reciprocity 117, 129
reflection, failure of critical 66
regulations (unintended consequences) 182–187
reimagining humanistic media studies 42–47; Balsamo, Anne 43; Cooley, Heidi Rae 42; Farman, Jason 46–47; Hoyt, Eric 44–45; Losh, Elizabeth 44; McPherson, Tara 45–46; Reimer, Bo 42–43; Vonderau, Patrick 43–44

Reimer, Bo 18; advice 237; attaining
resources 223; career advancement 233;
conceptual models 256; education,
mentoring, training 227–228; ethics
(unintended consequences) 187;
favorite part of applied media studies
84; finding collaborators 131; gaps in
field (of applied media) 259; impact
of collaboration 89; institutional
settings 95; meaning of applied media
studies 33–35; principal investigators
230; reimagining humanistic media
studies 42–43; time, energy, effort (of
collaboration) 92; translation between
fields 135–136
relationships 40, 54, 87, 89, 93, 129, 130,
136, 139, 148, 153, 170, 183, 196, 220;
cultivating 54–56; between fields 42
remediation 4
research: applying 192–202; commitments
92; improving processes 240;
information and communication
technology (ICT) 97–116; qualitative
social 194; roles of media researchers
214–215
researchers: interviews by community
106–110; interviews by visiting
103–106
Research Ethics Committee 195
Researchgate.com 132
resources 6, 15, 16, 17, 18, 19, 50, 61, 81,
87, 93, 96, 118, 120, 132, 185, 193, 219,
220, 253–262; acquiring 219; attaining
220–223; financial 219, 228–230
Rettberg, Jill Walker 119
Rice University 65, 68
Risam, Roopika 121, 122, 123, 124
The Roaring 'Twenties (Thompson) 57
Robertson, Mattie 33, 139, *172,* 174, 175,
176
Rossetti, Dante Gabriel 26
Rouch, Jean 110
Ruby on Rails 244

Saab, Joan 55
Sawchuk, Kim 254
Sayers, Jentery 254
Scalar 15, 16, 31, 48–59, 85, 122, 135,
186, 187, 221, 257; collaborations 58;
expect the unexpected 56–59; Terms of
Services 56, 57
scales, organizing across multiple 53–54
scanning servers 58
scholarly knowledge, translating 138

Schön, Donald 205
School of Arts and Communication
(Malmö University) 34
science, technology, engineering, and math
(STEM) 3, 9, 119, 123
science and technology studies (STS) 117
searching: AIDS Memorial Quilt *145;*
Lantern *see* Lantern
See the Future project 206–208
Selfie Course 83, 118, 119, 232
Selfie Researchers Network 118, 119, 132
Selfie Syllabus 119
Senft, Terri 118
September 11, 2001 8
servers, scanning 58
Shah, Nishant 190
shared languages 159–178, 161–169
Sidney Harman Academy for Polymathic:
Studies 223, 224
Silicon Valley 38
site visits 110–113
Slack 10
slack exchange *167*
Slavery at South Carolina College 1801–1865;
website 170
small data, qualitative analysis of 193
Smithsonian Museum Folk Life: Festival 148
Snickers, Pelle 37
sniffing out communication *199*
Snow, C.P. 5
social media 56, 65, 75
Software Studies 13
South Carolina College 161
South Carolina Educational TV: (SC-ETV)
139
Spitulnik, Debra 97
Spotify 18, 83, 197, *199,* 200
Srinivasan, Ramesh 99
stakeholders: collaborations 87; large
courses with 95
Stanford 118
Star, Susan Leigh 38, 255
Starosielski, Nicole 38, 256
State of Exception (Agamben) 123
STEAM 3, 9
Steinberg, Erwin 6, 7
Stockholm University 94
storyboard for "Medical Media Arts Lab" *70*
streaming 195, 196, 197, 199; architectures
198; business-to-business side of 18;
capacities 75; live 138; in political
campaigns 200; as solution to cultural
problems 197; Spotify 83 (*see also*
Spotify)

Streaming Heritage: Following Files in Digital Music Distribution (2014–2018) 196
Surfacing project (2014, 2015) 256
sustainability: architectures of 219–237; Lantern 248–250
Sweden 221, 223
Swedish Radio 197
Swedish Research Council: (Vetenskapsrådet) 18, 184, 198
Syllabi 117–124

Taylor, Toneisha 122
teams 16, 63, 69, 70, 91, 129, 131, 157, 160, 174, 175, 219, 227:Lantern *247*
technology 39; applying research 192–202; collaborations 51; development protocols 43; interventions 4; knowledge 6
Technology and Culture (Feibleman) 6
"Technology Is to Science as Female Is to Male" (Cowan) 6
television 6 *see also* film; video
tenure 230–233
Terms of Services (Scalar) 56, 57
terrorist attacks, September 11, 2001 8
Terry, Jennifer 52
Texas Medical Center (TMC) 62, 63 *see also* Medical Futures Lab
text-based data processing 24
textual deconstruction (Derrida) 7
Textual Encoding Initiative (TEI) 25
Theater of the Oppressed 47
theory: critical 93; film 41; reorientation of 44
third task *(tredje uppgiften)* 138
Thomas, Nicholas 193
Thompson, David 62
Thompson, Emily 57
Thompson, Nicholas 138
time, energy, effort (of collaboration) 91–94; Cooley, Heidi Rae 93; Farman, Jason 93; Hoyt, Eric 93–94; Losh, Elizabeth 91–92; McPherson, Tara 92–93; Parks, Lisa 91; Reimer, Bo 92; Vonderau, Patrick 91
Topgaard, Richard 206
Topol, Eric 69
trading zones 43
training 223–228 *see also* mentoring Tran, Tony 245
transdisciplinary collaborations 129–140

transformation 86; of cultural arrangements 35; of media production 10; urban environments 203–216
translation: between fields 133–140; practices of 4, 129; scholarly knowledge 138
translational media-making 129–140
Treatise and Method (Coleridge) 5
tredje uppgiften (third task) 138
Treichler, Paula 155
Triola, Mark 69
Trotter, Fred 69
Trump, Donald 120
Tsivian, Yuri 36
Tumblr 68
Twitter 56, 75, 200

unintended consequences 181–191
University of California 33
University of London 254
University of Maryland 131, 132
University of New South Wales 254
University of Pennsylvania 58
University of South Carolina 190, 220
University of Southern California (USC) 89, 95, 96, 131, 220
University of Texas at Austin (UTA) 240
University of Texas Medical School in Houston 62
University of the Arts in Berlin 36
University of Toronto 254
University of Victoria 254
University of Waterloo 253
University of Wisconsin 44
University of Wisconsin-Milwaukee 40
Unsworn Industries 207
urban environment transformations 203–216
user experience (UX) 162, *165, 166*
user interface (UI) 162, 167, 185
user navigation (UX) 185
U.S. State Department 221

Van Damm, Andy 148
Van Hise, President 44, 46
van Stam, Gertjan 100
Variety 239
Vartabedian, Bryan 65, 67, 68
Vectors 48, 49, 85, 89; collaborations 49–51; expect the unexpected 56–59; experimentation and process 51–53; external relationships, cultivating 54–56; finding collaborators 130; launch of 220; organizing across multiple scales 53–54

Vetenskapsrådet (Swedish Research
    Council) 184
video 31, 52, 209; across different contexts
    and fields 226; from Afghanistan and
    Iraq wars 52; collaborations 110;
    Condit, Cecelia 40; consumer video
    recorders 33; curriculums 118; design
    52; editing 85; games 13, 95; interviews
    103, 106, 108, 109; online channels 38;
    production of 68, 69, 225, 245; remote
    usage of 72; research 109; scholarly
    use of 40; sharing 75, 109; streaming
    capacities of 75; viewing 206; on
    YouTube 183 (see also YouTube)
Virtual AIDS Quilt digital image 146
Virtualpolitik (Losh) 37, 188
Virtual Quilt images 148
Viscomi, Joseph 25
vision statements, Medical Futures Lab 64
Vonderau, Patrick 18; advice 234–235;
    attaining resources 221; career
    advancement 231; change in roles 190;
    conceptual models 255–256; education,
    mentoring, training 225; ethics
    (unintended consequences) 184; favorite
    part of applied media studies 82–83;
    finding collaborators 132–133; gaps in
    field (of applied media) 259; impact of
    collaboration 88; institutional settings 94;
    meaning of applied media studies 35–37;
    principal investigators 229; reimagining
    humanistic media studies 43–44; time,
    energy, effort (of collaboration) 91;
    translation between fields 138

Ward One 33, 87, 93, 170–171, 185–190,
    221; collaboration with members
    171–176; cultivation time 139;
    meetings 173; user experience (UX)
    diagram 166 Ward One: Reconstructing
    Memory 175
Webby Awards 135
Wend, William Patrick 122
Weyeneth, Robert 170
"What Happened before YouTube"
    (Jenkins) 12
Where the Action Is: The Foundations of
    Embodied Interaction (Dourish) 49
Whitman, Walt 25
Whose Digital Village? (Srinivasan) 99
Williams, Chad 120, 124
Williams, Kidada 120
Williams, Mark 260
Willinsky, John 184
wireframes 166
Wisconsin Idea 44, 45, 237
witness journalism 138
Women Writers in English 1350–1850 25
Women Writers Online 25
Women Writers Project (WWP) 25
workshops, finding collaborators 132
World War II 6, 122

YouTube 56, 75, 183

Zambia 16, 17; Macha (fieldwork in)
    99–113; media fieldwork 97–116;
    mobile phones 105; Mudenda,
    Consider 106

For Product Safety Concerns and Information please contact our EU
representative GPSR@taylorandfrancis.com
Taylor & Francis Verlag GmbH, Kaufingerstraße 24, 80331 München, Germany

www.ingramcontent.com/pod-product-compliance
Lightning Source LLC
Chambersburg PA
CBHW071410050326
40689CB00010B/1819